God Versus Nature

God
Versus
Nature

THE CONFLICT BETWEEN
RELIGION AND SCIENCE
IN HISTORY

Frederick M. Seiler

Epigraph Books
Rhinebeck, New York

Paperback ISBN 978-1-951937-04-1
eBook ISBN 978-1-951937-05-8

Library of Congress Control Number 2020900030

Epigraph Books
22 East Market Street
Suite 304
Rhinebeck, NY 12572
(845) 876-4861
epigraphps.com

CONTENTS

PREFACE

S CIENCE AND RELIGION are both powerful forces in today's world. Science has given us an unprecedented level of understanding of our physical surroundings and even our own nature as living, thinking creatures. Science-based technology has thoroughly transformed the way we live in modern societies, providing an impressive array of life-giving and life-enhancing tools.

Religion guides the lives of the vast majority of the population of the earth. Even in relatively secular countries, the lives of most are strongly affected by ideas that come from religion. Whether the power of religion is increasing or decreasing, it is crucial for us to understand its nature and its relation to science.

My goal in writing this book has been to create an essentialized overview of the conflict between science and religion throughout history—an overview informed by the fundamental philosophic natures of both science and religion. Numerous thinkers have correctly identified the essential issue as the conflict between faith and reason; however, I disagree that this is the most *fundamental* philosophic issue.

This book was inspired by two famous books from the late nineteenth century: John William Draper's *History of the Conflict Between Religion and Science*, and Andrew Dickson White's *A History of the Warfare of Science with Theology in Christendom*. Both Draper and White correctly saw the fundamental relationship between science and religion to be one of conflict, but unfortunately, both books are flawed—most notably, they have numerous historical inaccuracies. For this reason, academic historians typically dismiss these accounts as historically worthless.

I am aware that my theme goes against the current trend among academic historians of science, who talk disparagingly about an outmoded "conflict thesis" and who refuse to look for broad patterns in history. But the fact is that there *are* broad patterns in history, and they can be found by those who are not mired in the myopia and complexity worship of academe.

Coming from an atheistic perspective, this book is directed primarily at the nonreligious reader who is curious about how and when science and religion have clashed. However, active-minded religious readers may also profit from the book, as an intellectual challenge.[i]

Given the basic conflict between science and religion, a close look at history reveals some puzzling facts:

- Science was born in a society that believed in many gods (Ancient Greece).
- Numerous scientific achievements were made in the very religious Islamic world.
- Modern science was born in a society dominated by Christianity (seventeenth-century Europe).
- Most scientists in history were religious.

How are we to make sense of these facts? How are we to relate them to the broader trajectory of the science/religion relationship from Ancient Greece to the present?

That is the subject of this book.

---

[i] For a good overall critique of religion, I recommend any of the following: *Atheism: The Case Against God*, by George H. Smith; *The End of Faith: Religion, Terror, and the Future of Reason*, by Sam Harris; *The God Delusion*, by Richard Dawkins.

S OMETIME IN THE near future, your child may pick up his textbook to learn about the famous scientist Galileo and his confrontation with the Church, and he may read:

> The reason Galileo came into conflict with the Church was because of his arrogance and his bad luck, not because of any supposed conflict between science and religion. Science and religion cannot conflict, and they are both valid guides.

This, of course, is not correct. Galileo's confrontation with the Church, properly understood, is a symbol of a fundamental conflict— the conflict between science and religion, between reason and faith. The idea that science and religion are fundamentally at odds has been called the "conflict thesis" by historians of science. This image of conflict has been accepted by many throughout the twentieth century.

But in recent decades, a different view has been accepted by the professionals responsible for understanding this event and others like it. According to many of today's historians of science, there is no basic conflict between science and religion, and there is no evidence for such a conflict in history:

- "A historical survey of the relationship between science and religion reveals that they cannot be seen either as natural allies or as natural enemies."[1]
- "Science and religion interactions in Europe . . . were extremely complex — sometimes mutually supportive, sometimes mutually antagonistic, and more often simultaneously supportive and

antagonistic, depending on what particular place one occupied within the spectrum of both religious and scientific attitudes, ideas, and practices."[2]

- "At different phases of their history, science and religion were not so much at war as largely independent, mutually encouraging, or even symbiotic."[3]
- "The idea that scientific and religious camps have historically been separate and antagonistic is rejected by all modern historians of science."[4]

The views of today's historians become part of tomorrow's schoolbooks, so the conventional Galileo story—the symbol of science/religion conflict—can't survive long.

These historians are missing something crucial and fundamental. This book has been written to address this lacuna. What is the proper way to understand the relationship between religion and science in history? That is addressed in the chapters that follow.

CHAPTER ONE

The Fundamental Natures of Religion and Science

T HE THREE PEASANT girls were out looking for firewood, which was urgently needed for cooking. In order to avoid accusations of stealing, they had walked out of town to an area of common land, where they could gather driftwood.

The youngest of the three was fourteen years old, and she was also the weakest. She suffered from several ailments, including asthma. Unable to keep up with the older girls, she stopped near a rock formation with a small cave at its base. At this point, according to her account:

> She saw a soft light coming from a niche in the grotto and a beautiful, smiling child in white who seemed to beckon to her. She was startled and instinctively reached for her rosary, but was unable to pick it up until the child produced one herself and began to make the sign of the cross; then she watched until the girl disappeared.[5]

The year was 1858, the peasant girl was named Bernadette Soubirous, and the place was Lourdes in southern France.

This was the first of eighteen visions. It was only Bernadette who saw and heard them, although others were present at many of them. One of the key visions occurred on the day of the Annunciation—March 25. Bernadette asked the vision who she was, and she received the answer: "I am the Immaculate Conception." The ghostly girl in white was apparently the Virgin Mary.

The adults and town officials were at first skeptical of Bernadette's claims, and Church authorities were reluctant to grant them legitimacy. But Bernadette never showed the slightest inclination to recant her claims; she seemed truly sincere. Many local women became utterly convinced that the Mother of God had chosen their obscure town of Lourdes to reveal herself to humanity. Their excitement proved contagious:

> By the time Bernadette died in her convent in Nevers, hundreds of thousands were going to the shrine every year, as rich and poor alike, women and then men, concluded that she had indeed been granted a genuine encounter with the Virgin. . . . In 1908 more than one and a half million pilgrims came to commemorate the fiftieth anniversary of the apparitions; a similar number came in 1950 when the Assumption of the Virgin was declared a dogma of the Catholic Church; almost five million were coming every year in the early 1990s, now arriving from all over the world on package charter flights.[6]

While the story of Lourdes is unique in many ways, it does highlight several important components of religion. A girl's extraordinary claims led to the widespread belief that she had experienced miraculous visions. These were visions of a supernatural being—a being that is not part of the natural world. Why were so many people so eager to believe the claims? Because they wanted the claims to be true; the belief made their world seem to be a thrilling place, an important place; the claims confirmed their beliefs in the miracles of Christianity. Above all, because they had faith.

* * *

The king's physician was brimming with curiosity; he wanted to understand how the human heart really works. But his contemporaries considered this an impossible dream. The heart is so complex and moves so quickly (and stops so soon after an animal is surgically opened) that it seemed impossible for the human eye to capture what was really happening. The physician was William Harvey, physician to King James I of England.

According to the writings of Galen—the revered Greek physician—blood is continuously generated in the liver from the food we eat, from where it is pumped out by the heart to the rest of the body. The blood never returns to the heart; it is simply consumed by the body.

Harvey had tremendous respect for Galen, but he knew that Galen made mistakes. The Renaissance anatomist Andreas Vesalius had found clear evidence that Galen's view of blood was flawed. Unfortunately, no one, including Vesalius, had proposed an alternative system to Galen's.

Several physicians had recently discovered that blood leaves the heart from the right ventricle, travels to the lungs (where its color is transformed from a dull red to a bright red) and then travels to the left atrium of the heart. This loop from the heart to the lungs and back to the heart was called the pulmonary circulation, and this was considered the only closed loop involving the heart.

Vesalius and others had discovered that inside the veins there are valves, which seem to allow blood to flow in only one direction. Vesalius thought the valves merely strengthen the veins, and others thought that the valves are gravity-directed, to stop blood from pooling up in the lower part of the body. But Harvey observed that the valves in the jugular vein (from the heart to the head) point downwards (in the same direction as gravity), so the commonality among the valves is that they always point toward the heart.

Harvey then performed an ingenious set of experiments with ligatures—experiments which exploited the fact that in the human arm the arteries lie deeper than the veins, so a moderately tight ligature blocks the veins, and an extremely tight ligature blocks both the arteries and the veins:

> If we begin with a tight ligature, we find that the hand becomes cold but remains properly coloured. If we slacken the ligature somewhat, then the hand becomes flushed and the veins stand out and become distended. If we then release the ligature altogether,

we release the blood in the veins and the arm returns to normal. Harvey concluded from this that blood was carried out in the arteries. With a tight ligature the arteries are cut off as well as the veins, so the arm does not swell and there is no pulse. With the lighter ligature the arteries still carry blood out, but as the veins are restricted they cannot carry blood back and so there is a swelling of the veins. So blood is carried out by the arteries, finds its way into the veins, and is carried by the veins back to the heart.[7]

Harvey also performed a similar set of experiments using a finger to apply pressure on a vein to block it at certain points. The results of these experiments all confirmed the idea that blood flows back to the heart through the veins. But Harvey's most famous and persuasive argument for the circulation of the blood involved his use of quantitative reasoning:

If the human heart contained two ounces of blood (an observation from cadavers) and made about sixty-five beats per minute, then in one minute it pumped about eight pounds of blood. This amount multiplied by the minutes in a day gave a fantastic quantity of blood, far too much for the body to produce rapidly from food eaten. Harvey further supported these speculations with experiments on live sheep. Severing a sheep's main artery he collected and measured the blood expelled in a unit of time. It became obvious to him that blood circulated in a closed system.[8]

Harvey's theory brought together his knowledge of the pulmonary circulation, the valves in the veins, his experimental results with ligatures, and his numerical calculation of the rate of blood flow through the heart. But how exactly did the blood get from the arteries to the veins? Harvey hypothesized the existence of minute passages from the arteries to the veins, but nobody was able to observe them during his lifetime. Several decades later, microscopist Marcello Malpighi finally observed the capillaries, the missing element in Harvey's theory. Finally, Harvey's theory was conclusively proven.

The story of William Harvey's discovery of the circulation of the blood is an inspiring example of science at its best. For our purposes, note what this story highlights about the nature of science. William Harvey was motivated by an intense curiosity about the natural world. He used a combination of careful, systematic observation and logical reasoning based on his observations (confirmable by others) to arrive at a theoretical conclusion that integrates a wide variety of facts. William Harvey used *reason* in order to understand *reality*.

The Nature of Religion

THE PRECEDING EXAMPLES have highlighted several aspects of religion and science. In this section and in Chapter Four, we'll consider a more philosophical analysis of the two, in order to better understand their relation to each other. This will help us, in the rest of the book, to understand the ways in which the two have been related in history.

The Oxford English Dictionary defines religion as "Belief in or acknowledgement of some superhuman power or powers (esp. a god or gods) which is typically manifested in obedience, reverence, and worship." A religion is a set of views about the supernatural origins, workings, and purposes of reality, and about what this implies for the living of our lives.[9]

More broadly, a religion is a type of *philosophy*; it is a systematic set of answers to the basic questions faced by all human beings throughout history:

- Where am I? (What kind of world do I live in?)
- Who am I? (What kind of a being am I?)
- How do I know it? (How do I obtain knowledge about the world?)
- What should I do? (How should I live my life?)

Since these questions are universal, the human need for philosophy is universal. The dominant form of philosophy in history has been religion, making religion virtually ubiquitous. The development of non-religious philosophic ideas has been relatively rare in history, with Ancient Greece as a significant exception.

According to religion, the world around us—the world that we see, hear, touch, smell, and taste—is not the only world. It is not even the most important world. There are important, powerful things beyond nature—that are "supernatural"—entities like gods, angels, and demons, and places like heaven and hell. According to western religions such as Christianity and Islam, after we die, we live in an afterlife for the rest of time. Our limited time on earth is nothing compared to eternity in the other world. So should we care very much about this world? Religion's answer is clear: No.

Belief in another, higher world and an eternal afterlife logically leads to denying the central importance of the world that we see, hear, touch, smell, and taste.

The most important thing is discovering how to best prepare for the next world. We need to get in touch with the powers of the next world, to find out what we should be doing. According to religion, the most important things are not available to the senses or direct reasoning from them; this premise gives rise to a key religious concept: faith. Faith is belief in the absence of evidence or in contradiction to evidence. According to the biblical character Paul, faith is "the assurance of things hoped for, the conviction of things not seen" (Heb. 11:1). Faith is a key religious virtue. It is explicitly stressed in western religions such as Christianity, but it plays a crucial role in *all* religions. According to philosopher Leonard Peikoff, "'Faith' names the method of religion, the essence of its epistemology." *Epistemology* is the philosophy of knowledge, the study of how we know what we know. A closely-related branch of philosophy is *metaphysics* (or *ontology*), which is the study of the basic nature of existence or reality. Peikoff argues that the metaphysics of religion is belief in "some higher unseen reality" or power:

> According to religion, this supernatural power is the essence of the universe and the source of all value. It constitutes the realm of true reality and of absolute perfection. By contrast, the world around us is viewed as only semi-real and as inherently imperfect, even corrupt, in any event metaphysically unimportant.

According to most religions, this life is a mere episode in the soul's journey to its ultimate fulfillment, which involves leaving behind earthly things in order to unite with Deity. As a pamphlet issued by a Catholic study group expresses this point: Man "cannot achieve perfection or true happiness in this life here on earth. He can only achieve this in the eternity of the next life after death. . . . Therefore . . . what a person has or lacks in terms of worldly possessions, privileges or advantages is not important."[10]

A concept closely related to religion is "mysticism." Mysticism is the philosophic view that we can obtain knowledge from means other than reason and sense perception. Mystics have proposed such means as faith, feelings, intuition, revelation, or "just knowing." Mysticism is a key component of religion.

The Nature of Science

SCIENCE IS THE systematic study of natural phenomena in order to grasp the natures of entities and their actions. It involves the systematic use of observation and logical reasoning to develop theories that explain the world in which we live. As one scientist summarizes:

> The scientific method involves the observation of phenomena or events in the real world, the statement of a problem, some reflection and deduction on the observed facts and their possible causes and effects, the formation of a hypothesis, the testing of the hypothesis (experimentation or prediction), and—when the tests repeatedly confirm the hypothesis—the erection of a theory.[11]

The resulting knowledge is a powerful tool for improving human life, although the full extent of science's usefulness was not evident until the Industrial Revolution. Before the eighteenth century, it was commonly thought that the purpose of science was knowledge for its own sake.

The Scientific Revolution of the seventeenth century led to the systematic use of experimentation and mathematics, which have become central to what we call science today.

In the example of William Harvey, we saw him doing a wide range of observations and experiments. Harvey even used mathematics—in the form of a crude numerical measurement. But the real power of mathematics was in the fields of astronomy and physics, where the discoveries of Galileo, Kepler, and others led to the brilliant synthesis of Newton's theories and countless other discoveries.

Given the critical importance of the Scientific Revolution, did science really exist before it? Yes, it did. As we shall see in the next chapter, the Ancient Greeks achieved the basic scientific mindset and made a number of important scientific discoveries.

The Ancient Greek Rejection of Mysticism Leads to the Birth of Science

Pre-Greek Civilizations as Mystical

As we discussed in Chapter One, religion addresses the genuine human need for an integrated set of answers to the basic questions: Where am I? Who am I? How do I know it? What should I do?

So it is not surprising to find that religion has been ubiquitous in history. In primitive religions across the globe, gods are everywhere, and observed natural events are explained by their actions. Thunder is caused by a sky god, such as the Germanic Thor; flooding is caused by a river god, such as the Hindu Ganga; earthquakes are caused by an earth god, such as the Māori Rūaumoko.

Consider how a child (primitive or modern) develops cognitively. He first grasps cause and effect from his own actions. He closes his eyes, and the world disappears. He drinks water, and his thirst is quenched. He releases a stick, and it falls. He acts, perceives the result of his action, and grasps the connection. He sees other people acting and he sees the results of their actions. He may easily overgeneralize from this and conclude that all things have minds like his, that cause their behavior. Trees have minds that cause them to turn their branches toward the sun; rivers have minds that cause them to flood or to dry up:

When they turn to cause and effect in the external world, primitives (and children left to themselves) typically continue to interpret the causal processes they perceive on the model of their early personal experience. Projecting their own pattern outward, they think of causal agents in the external world, some or all, as being personal entities, and they construe causation as the expression of inner desires or intentions. The obvious historical example of such anthropomorphism is the animist idea that even inanimate entities are ensouled things who act with ends in view.[12]

This is a simple form of the religious view that *a mind is the source of causality*, a form that typically only occurs in primitive societies.

* * *

The earliest records we have which may in some sense be called scientific come from the ancient cultures of Egypt and Mesopotamia, which both go back thousands of years before the emergence of Greece. Both of these cultures made cognitive advances which helped the Greeks to develop science, but we will see that neither of these possessed science itself, because neither had a scientific mindset.

Egypt

The pyramids are dramatic evidence of the Egyptians' skill in large-scale construction and their ability to perform and apply geometric calculations. The Egyptians developed a simple numerical notation that had ten symbols but was not position-based like our modern system. Using this system, they were able to solve simple problems of area, weight, and volume.

The Egyptians also made simple astronomical calculations related to maintaining the calendar and aligning temples. The north face of the Great Pyramid at Giza is precisely aligned (within 1/20 of a degree) to the celestial north pole, the point around which the rest of the stars appear to rotate. Their calendar served both agricultural purposes (when to plant the seeds) and religious purposes (when to perform ceremonies). Their skill in embalming revealed a rudimentary knowledge of human anatomy, and papyri have been found recording the progression of symptoms for diseases.

But perhaps the most notable aspect of Egyptian civilization was its changelessness. Early in their history, Egyptians made a few key advances in mathematics, medicine, and calendrical astronomy, but over the course of two thousand years, there were virtually no discoveries or advances in these fields. This extreme conservatism makes sense in light of the otherworldliness of Egyptian culture. The Egyptians had an elaborate and centralized state religion. The pyramids themselves are monuments to the Egyptian obsession with the world beyond death. Given the absolute importance of the other world, the Egyptians had little motivation for studying this world or improving their life on Earth.

Mesopotamia

The Babylonians developed a complex numerical notation that was both decimal and sexagesimal (based on powers of 60). Babylonians recorded "word problems" in which specific unknown quantities had to be calculated from known quantities. They compiled tables of multiplications, reciprocals, squares, and square roots, as aids to their calculations.

Babylonian priests compiled extensive records of the motions of the celestial bodies—records that would later be used by Greek astronomers. These priests made accurate mathematical calculations for planetary positions and times, and compiled complex tables for prediction of celestial events such as eclipses and lunar phases. They used these astronomical calculations to create a calendar based on the lunar cycle but also related to the solar year. The maintenance of an accurate calendar was crucial to the Babylonians for the same reasons as for the Egyptians: for agricultural and religious purposes.

The Babylonians' medical practices, like those of the Egyptians, were strongly tied to their religion. According to historian Peter Whitfield:

> The overriding belief was that disease was caused by divine disfavour, or demonic possession. This did not mean that the sick were not treated, or were thought to be beyond human help, but

it did mean that no functional theory of disease emerged, and treatment was directed entirely to the symptoms. Several different professional groups were involved with medicine: the diviner, the exorcist and the physician, and the practice of hepatoscopy — the examination of the liver of sacrificed animals — was the most important single means of diagnosis.[13]

Ancient Mesopotamian and Egyptian mathematics did not rise to the level of a science. Despite a proficiency in certain mathematical operations, there was no conceptual-level mathematics with general propositions, theorems, or proofs. Moreover, according to Whitfield:

> Despite centuries of meticulous sky-watching, there is no theory of the shape or mechanism of the cosmos; an intimate knowledge of the human body produced no theory of the functions of its organs, or even the concept of functional disease; the ubiquity of physical working against mass and weight did not evoke any language of physics. Their philosophy of nature led them to explain the world they saw by creating a parallel world that was unseen, a world whose laws integrated the realms of gods and men, life and death.[14]

For both of these ancient civilizations, religion was the central focus of their abstract conceptual thought.

Ancient Greece as the First Non-Mystical Culture in History

THE ANCIENT GREEKS had an unusual religion. They believed in the Homeric gods, but their gods were conceived of as powerful humans. As author Thomas Cahill points out: "What is so striking about the Homeric gods . . . is their lack of godliness."[15] These gods could perform morally wrong actions, and they did. There was no organized religion; there were no central authorities and no canonical religious texts. The gods had limits on their power; they had to work within the limits of reality. It was widely accepted that the world had always existed and that it did not need a god to make it come into existence.

The most mystical aspect of Greek culture was the belief in oracles, which were often consulted for major decisions. But oracles were considered fallible, and the Greek philosophers did not take them seriously.

In both Babylon and Egypt, learning and specialized study of the observed world were only done by priests. In contrast to this, the Greek scientists were private citizens and teachers, who had no courtly role or royal patrons to serve. According to philosopher Leonard Peikoff:

> Ancient Greece was not a religious civilization. . . . The gods of Mount Olympus were like a race of elder brothers to man, mischievous brothers with rather limited powers. . . . They did not create the universe or shape its laws or leave any message of revelations or demand a life of sacrifice. Nor were they taken very seriously by the leading voices of the culture, such as Plato or Aristotle. From start to finish, the Greek thinkers recognized no sacred texts, no infallible priesthood, no intellectual authority beyond the human mind; they allowed no room for faith. . . . Men generally dismissed or downplayed the supernatural. . . . There was a shadowy belief in immortality, but the dominant attitude to it was summed up by Homer, who has Achilles declare that he would rather be a slave on earth than "bear sway among all the dead that be departed."[16]

Ancient Greeks held that our central problem is the achievement of well-being in this world. Further, they held that we have the intellectual and moral capacities to solve this problem and to fashion a good life for ourselves by our own efforts. Whitfield has pointed out how the "shallowness" of Greek religion was responsible for the birth of Greek philosophy and science:

> Having no profound faith in their gods, their intellect was free to explore realms of thought unknown to other ancient civilizations. The object of their study was *physis* — nature in its widest sense — and they thus became the first *physikoi* — physicists. The very word which they gave to this world, *kosmos*, meant in its original sense 'order', and this intellectual adventure seems to have

sprung from the conviction that the world was an ordered system, which would yield to rational investigation.[17]

Historian of science G. E. R. Lloyd argues that the Milesian philosophers such as Thales made two significant intellectual achievements: "the discovery of nature" and "the practice of rational criticism and debate." By the "discovery of nature" he means

the appreciation of the distinction between the "natural" and the "supernatural," that is the recognition that natural phenomena are not the products of random or arbitrary influences, but regular and governed by determinable sequences of cause and effect. Many of the ideas attributed to the Milesians are strongly reminiscent of earlier myths, but they differ from mythical accounts in that they omit any reference to supernatural forces.... [T]he supernatural plays no part in their explanations.[18]

Lloyd points out, as an example, the theory of earthquakes which is attributed to Thales. Thales believed that the earth rests on water; he proposed that earthquakes are caused by the earth being rocked by wave-tremors in the water. This account was extremely simple, but it made no reference to Poseidon or any other god.

The pre-Socratic natural philosophers such as Thales and Anaximenes were looking for a single natural substance (or a few substances) that could account for all of the substances of the natural world. While the substances that they proposed (water, air, fire, etc.) were naïve, their search for the "one in the many" was an important way to make sense of their experience. This type of search led Leucippus and Democritus to hold that everything consists of solid impenetrable atoms; and Empedocles to conclude that everything consists of four elements: earth, air, water, and fire.

For another example of the Greek naturalistic turn, we can look at the extensive body of medical writings known as the "Hippocratic corpus." Associated with the fifth-century physician Hippocrates of Cos and his followers, these writings cover a wide variety of medical subjects.

Generally, in the Hippocratic writings, intervention by the gods was ruled out as a direct cause of disease. A Hippocratic work titled *Airs, Waters, Places* states that "Each disease has a nature of its own, and none arises without its natural cause."[19] An excellent example of this is the way epilepsy was treated. Epilepsy had always been explained by reference to the gods, and it was known as the "sacred disease." But the Hippocratic author of *On the Sacred Disease* rejected this view, and started the work as follows:

> I do not believe that the "Sacred Disease" is any more divine or sacred than any other disease but, on the contrary, has specific characteristics and a definite cause. Nevertheless, because it is completely different from other diseases, it has been regarded as a divine visitation by those who, being only human, view it with ignorance and astonishment.[20]

Plato Creates a Philosophy Based on the Primacy of Abstract Forms

PLATO AND ARISTOTLE—the two philosophical giants of Greece—created two dramatically different philosophical systems, especially in terms of their metaphysics and epistemologies.[21] These led to two different attitudes toward the study of the natural world.

Plato (428–348 B.C.[22]) taught that the world we experience is not fully real, that the individual physical objects we see are imperfect, corrupted reflections of those in a higher reality: the world of the Forms. In Plato's *Republic*, in order to explain his view of knowledge, he presents his famous myth of the cave:

> Imagine human beings living in an underground cave-like dwelling. . . . They've been there since childhood, fixed in the same place, with their necks and legs fettered, able to see only in front of them, because their bonds prevent them from turning their heads around. Light is provided by a fire burning far above and behind them. Also behind them, but on higher ground, there is a path stretching between them and the fire. Imagine that along

this path a low wall has been built, like the screen in front of pup-
peteers above which they show their puppets. . . . Then also imagine that there are people along the wall, carrying all kinds of artifacts that project above it—statues of people and other animals,
made out of stone, wood, and every material.[23]

In this case, the prisoners would naturally believe that the only reality is the shadows that they see. But, Plato then asks, what if a prisoner
were freed from his bonds and compelled to stand up and turn around?
"He would be pained and dazzled and unable to see the things whose
shadows he'd seen before." Eventually he would grasp that the artifacts
were more real than the shadows he'd believed were the only true reality.

Then if he were taken out of the cave into the sunlight, he would take
more time to adjust to the light. Finally, he would see the sun: "And at
this point he would infer and conclude that the sun provides the sea-
sons and the years, governs everything in the visible world, and is in
some way the cause of all the things that he used to see." If he then
reminded himself "of his first dwelling place, his fellow prisoners, and
what passed for wisdom there," he would "count himself happy for the
change and pity the others."

In this myth, the flickering shadows represent the observable world
that we live in. The objects that cast the shadows are the Forms, and the
sun is the Form of the Good, which is the ultimate source and sustainer
for everything else. Peikoff summarizes Plato's metaphysics:

Reality . . . is not the physical world of Nature, not the imperfect
sensible world of concrete men and things, with its ceaseless flux
of contradictions. . . . Concretes, including individuals, are merely
appearance, the distorted and ultimately unreal shadows of a
higher, non-material, truly real dimension, the world of Forms or
abstractions, which is motionless, logical, and perfect.

The Forms are hierarchical, culminating in the pinnacle of reality,
the ineffable Form of the Good. . . . This Form is what gives unity
and meaning to the universe. It is the fundamental fact, from
which all the lower Forms (and thus all the shadows) derive and

on which they depend. And it is the fundamental value, toward which all things aspire.[24]

Plato created his theory of Forms as a solution to the problem of universals, which can be stated as follows: Concepts (such as man, horse, justice, beauty) are abstract and general. The concept "man" somehow refers to all men, everywhere, past, present, and future. But what exactly does "man" refer to? All the men we see differ in all of their attributes: they differ in intelligence, character, strength, health, size, and every other attribute. Plato's answer to this problem was that there is a single, eternal, perfect Form of "man" which exists in a world of Forms, and that all individual men somehow "participate" in that one Form; they are, in effect, shifting reflections of the Form of "man."

So how can we come to know the Forms—the true reality—according to Plato? Plato holds that in a previous life we were in touch with the Forms, but on being born into this world we have forgotten them. In order to get back to them we must somehow "recollect" them. Initially, some items in the visible world may allow us to start our recollection, but we must soon turn away from the shadows that surround us and train our minds to deal in eternal timeless truths, such as the truths of mathematics. This is why Plato was an ardent supporter of the study of geometry.

Because of his support for the study of geometry, Plato has often been esteemed as pro-science. But Plato's interest in geometry came primarily from its perceived ability to train the mind to ignore physical reality and come into contact with the Forms. Plato did not see mathematics as a tool for better understanding the physical objects around us: "If we are ever to have pure knowledge, we must escape from the body and observe matters in themselves with the soul by itself."[25] In the *Republic*, Plato derogates the study of the starry sky as being too physical, too this-worldly, concluding with the advice: "Let's study astronomy by means of problems, as we do in geometry, and leave the things in the sky alone."[26]

Peikoff summarizes Plato's advice: "Do not play the pointless game of the cave dwellers (such as empirical scientists), who look for order in the chaos of shadows. Do not *look*, but think, by turning your attention to the *a priori* ideas comprising reality."[27]

In spite of Plato's dim view of worldly knowledge, he did create a cosmology. In Plato's *Timaeus*, he presents an account of the creation and structure of the world which he viewed not as actual knowledge of the physical world—which he regarded as impossible—but as a "likely story." In the account, a master craftsman, the Demiurge, has created the world by taking pre-existing chaotic material and trying to shape it according to the Forms. Of course, being naturally chaotic, the material did not fully accept the Forms. The *Timaeus* presents Plato's views on astronomy, cosmology, light and color, the elements, and human physiology. The *Timaeus* would later be translated into Latin by Cicero, and it would become influential in the Middle Ages.

Plato was a Greek anomaly; he was too mystical for most Greeks. He held that in order to gain true knowledge, we have to disregard the observable world, and turn our glance inward, into our own minds. This premise would grow even more prominent in the Neoplatonists, and later Augustine, who would hold that the Forms are ideas in the mind of God.

Aristotle Creates a Philosophy Based on this World

ARISTOTLE (384–322 B.C.), IN CONTRAST to his teacher Plato, held that the objects of the physical world are fully real and thus worthy of study. Knowledge, he held, is gained not from groundless deductions and intuition, but ultimately from induction—from the study of the concrete, particular things we see, touch, smell, and hear around us. According to Aristotle, this world is *real* and worth studying for the knowledge it brings: from the beautiful stars in the night sky to the minutest details of slimy sea creatures. Aristotle held that "Every realm of nature is marvelous."[28] Not only was Aristotle's philosophy supportive of science, Aristotle was a scientist himself, and his impressive work in biology demonstrates a mind fully in touch with and keen on understanding the physical world.

Aristotle's epistemology was very different from Plato's. Aristotle rejected the idea of a separate, higher realm of Forms. He held that all knowledge is ultimately grounded in perceptual observations and that forms are in the objects we experience. When we look at a group of similar objects, our minds are able to use the faculty of abstraction to grasp the common form of the objects. This is how we think abstractly; but at the same time, we understand that abstractions cannot exist apart from concrete objects.

Immersed in the Greek culture that valued discussion and debate, Aristotle studied the many arguments that different thinkers gave for their views. Some arguments were clearly good, and some were clearly not. He set out to determine what all valid arguments had in common. This led him to develop an entirely new field of knowledge: the science of logic. Key to this science was the Law of Non-Contradiction: There are no contradictions in reality. The same thing cannot be both true and false in the same respect at the same time. For Aristotle, according to Peikoff: "The Law of Non-Contradiction is not a 'pure' tool, unsullied by sense, but a plain description of observed fact. This law, he says, is an absolute, true of this or any world: it is true, in his famous phrase, of 'being qua being,' which is why compliance with it is a requirement of valid thought."[29]

Implicit in the *law of non-contradiction*, but not explicitly stated by Aristotle, is the *law of identity*, which was stated earlier by Parmenides: A thing is what it is; it is not what it is not.

Whereas Plato held that true reality—the realm of the Forms—is eternal and changeless, Aristotle opened his eyes and looked around him. He saw animals running, plants growing, clouds drifting, celestial bodies moving, and people doing a wide variety of things. Aristotle saw that change was undeniable, and he developed conceptual tools to understand the changes around him. Particularly powerful were his concepts of the potential and the actual. Earlier Greek thinkers had seen change as contradictory, but Aristotle simply saw it as involving a potential becoming actualized, given the appropriate conditions. An acorn is a potential oak tree. A boy is a potential man. A pile of stones is a potential house.

Aristotle posited four types of cause (or explanation): material, formal, efficient, and final. Consider a house. The material cause is the wood and bricks from which it was constructed. The formal cause is the shape or structure of the house. The efficient cause is the actions of the builders in constructing the house. And the final cause is the reason for which the house was built—to provide shelter from the elements.

All physical things, or substances, are composed of two parts, which Aristotle called matter and form. The matter is the stuff, and the form is the structure. Whereas in the mind we may mentally separate the two aspects, in reality the two are inseparable. In reality there is no such thing as pure matter without form, or pure form without matter. Whereas Plato saw the soul as something immaterial that was trapped in the physical body, Aristotle saw the soul as simply the *form* of the body.

Aristotle's observational skills are most evident in his zoological works; he was a great observer of living animals. As described by Aristotle scholar Jonathan Barnes, Aristotle's *History of Animals*

> discusses in detail the parts of animals, both external and internal; the different stuffs—blood, bone, hair, and the rest—of which animal bodies are constructed; the various modes of reproduction found among animals; their diet, habitat, and behavior. . . . And he is particularly informed and particularly informative about marine animals—fish, crustacea, cephalopods, testacea. [The *History of Animals*] ranges from man to the cheese-mite, from the European bison to the Mediterranean oyster. Every species of animal known to the Greeks is noticed; most species are given detailed descriptions; in some cases Aristotle's accounts are both long and accurate.[30]

Aristotle's biological research was focused almost exclusively on zoology; however, he inspired his student and successor Theophrastus to do similar, detailed studies of botany.

Being primarily interested in living things, Aristotle developed a teleological (goal-directed) view of the world, but not in a way that involved anything mystical or religious, or inanimate entities having conscious goals. For example, he held that plants move their leaves in order to absorb sunlight, and rocks fall to earth (and water flows downhill) to get closer to their natural place, but he did not hold that plants or rocks set conscious goals.

Aristotle's cosmology came from theories of earlier Greek thinkers, integrated with Aristotle's own observations and his own philosophy. He viewed the earth as a sphere fixed at the center of the cosmos, with an atmosphere extending up near the moon. The moon and other celestial bodies are embedded in spherical orbs that rotate uniformly. Their motion is caused in some sense by a self-conscious "unmoved mover," which is the closest Aristotle came to a concept of God. (Aristotle's unmoved mover is not aware of the rest of the universe, including us.) The sub-lunar world is made up of four substances, which were (from Empedocles) earth, water, air, and fire. Each of these has its natural place, which it seeks out. Earth's natural place is the center of the cosmos, so when we drop a rock, it falls toward the center of the earth. These four substances, in various combinations, make up the numerous types of substances we experience.

Most of Aristotle's physics and cosmology is now known to be incorrect; however, given the relatively primitive state of knowledge of Aristotle's time, it was a brilliant integration that explained a wide variety of phenomena. In order to fully appreciate Aristotle's scientific achievement, we must pay close attention to his underlying philosophy, especially his empirical view of knowledge. As Barnes notes:

> Our modern notion of scientific method is thoroughly Aristotelian. Scientific empiricism—the idea that abstract argument must be subordinate to factual evidence, that theory is to be judged before the strict tribunal of observation—now seems a commonplace; but it was not always so, and it is largely due to Aristotle that we understand science to be an empirical pursuit.[31]

Peikoff argues that Aristotle made a fundamental advance in developing the law of cause and effect:

> The meaning of Aristotle's theory . . . is that each entity has a

certain nature and, if not disturbed, acts only in accordance with that nature. . . . In Aristotle's terminology, each entity has a definite, limited potentiality to actualize, and must therefore act—move—accordingly. When the actuality . . . has been reached, the relevant motion ceases.

The enduring essence of this view, long accepted by scientists who dropped the idea of natural strivings, is that the world is ruled by natural law, so that there are no chance events and no miracles. . . . Even though Aristotle's statements on this topic are not always consistent, his theory is the first attempt in Western history to formulate, validate, and apply a universal law of Cause and Effect, and to do so in purely secular terms.[32]

The Scientific Achievements of the Greeks

BECAUSE OF THEIR nature-centered approach to knowledge, the Greeks made numerous scientific advances, in fields such as medicine, optics, mathematics, and astronomy.

Medicine

We've seen how the Hippocratic works are predominantly naturalistic in their discussions of health and disease. One prominent naturalistic theory, the theory of the four humors, was first proposed in these writings. According to this theory, there are four essential fluids in the body— blood, phlegm, yellow bile, and black bile, which map to the qualities hot, cold, wet, and dry—and any imbalance in these fluids leads to pain and disease. When possible, according to these writings, the physician should attempt to bring these humors back into the proper proportions.

Another significant development in Hippocratic medicine was the recognition of the value of elaborate clinical descriptions of diseased or injured patients, including their symptoms and behavior. Some of these writings present extensively detailed reports on particular patients, recording how their condition changed over time. These reports could help physicians predict the course of disease for future patients.

After the Hippocratic writings, the next major developments in Greek medicine came during the Hellenistic period (after 300 B.C.), in the Museum of Alexandria. Officially a temple to the Muses, and connected to the great Library of Alexandria, this was in actuality more of a research center, where prominent thinkers came to study and discuss philosophy, mathematics, astronomy, medicine, and other subjects. It was the greatest research center of antiquity.

There is evidence that several individuals associated with Aristotle's Lyceum (after Aristotle's death in 322 B.C.) were influential in founding or contributing to the Museum of Alexandria.[33] This connection is not surprising since Aristotle was the greatest ancient proponent of the study of the natural world.

During the early Hellenistic period, an important field of research at the Museum was anatomy and physiology. The most prominent leaders of this program were Herophilus of Chalcedon (330–260 B.C.) and Erasistratus of Ceos (c. 315–240 B.C.). They performed human dissections and vivisections of living animals (and perhaps even of condemned criminals) and put forth theories of physiology.

Herophilus's anatomical investigations led him to numerous discoveries. He identified the heart's main chambers and the blood vessels going to and from them. He identified the brain (along with its main components) as the center of the nervous system. He distinguished between sensory and motor nerves. He dissected the eye and described its membranes. He identified the ovaries of the female body and compared them to the testes of the male. He also coined many anatomical terms that remain with us today; these include *duodenum*, which comes from the Greek words for twelve and fingers. (Using his hands to measure its length, on average it came to the width of twelve fingers side by side.)

The culmination of Greek medicine was achieved with the work of Galen of Pergamum (A.D. 129–c. 216). Galen studied and practiced medicine in Pergamum and Alexandria, working for a while as a surgeon to gladiators. He later went to Rome and acquired a reputation as a great physician, eventually becoming personal physician to Roman emperor Marcus Aurelius and later emperors.

Galen was heavily influenced by the Hippocratic corpus as well as the writings of Herophilus and Erasistratus. From the Hippocratic corpus he took the idea of the four humors as the essential constituents of the body. Galen added the idea that a particular organ could be diseased, from its own imbalance of the humors.

Due to Roman prohibitions, Galen was not permitted to dissect humans, but he dissected a variety of animals, including monkeys. He became quite skilled at dissection and even wrote a how-to guide on the subject. His writings on the human body were extensive, covering anatomy, physiology, diseases, and treatments. As in some of the Hippocratic works, he did extensive observations of his patients.

Galen created a physiological system in which the heart is the source of the arteries, and pushes arterial blood out to the rest of the body. The liver produces venous blood, which also goes out to the rest of the body. The blood is consumed by the body; it does not flow in a closed loop. Although we now know Galen's system to be largely incorrect, it did make sense of his observations. Historian of science David Lindberg summarizes Galen's accomplishments:

> Galen's medical system proved exceedingly persuasive, dominating medical thought and teaching throughout the Middle Ages and into the early modern period. Part of its persuasive appeal lay in its comprehensiveness. Galen addressed all the major medical issues of the day. He could be both practical, as in his pharmacology, and theoretical, as in his physiology. He was philosophically informed and methodologically sophisticated. His work embodied the best of Greek pathological and therapeutic theory. It contained an impressive account of human anatomy and a brilliant synthesis of Greek physiological thought. In short, Galen offered a complete medical philosophy, which made excellent sense of the phenomena of health, disease, and healing.[34]

Mathematics, Mechanics, and Optics

Whereas Egyptian and Babylonian mathematical writings generally discuss specific problems, such as how to calculate the area of a particular field with particular dimensions, the Greeks turned their focus to more theoretical issues, such as how to get the general solution for the area of any triangle, starting with fundamental definitions, axioms, and postulates. Consider the Pythagorean theorem. The Babylonians had recorded many Pythagorean triples such as (3, 4, 5), where $3^2 + 4^2 = 5^2$, but it wasn't until the Greeks (and most probably the Pythagoreans) that this theorem was proved for the general case of all right triangles.

One of the most famous Greek mathematical works is Euclid's *Elements*, which was written around 300 B.C. (probably in Alexandria). Euclid started with the works of earlier Greek geometers and integrated their discoveries into a deductive system of axioms, definitions, demonstrations, and proofs. *Elements* became "the most enduring and widely-studied secular book in the western world, dominating the teaching of mathematics for more than two thousand years."[35]

Another mathematical giant of antiquity was the creative genius Archimedes (c. 287–212 B.C.). Archimedes is known to have visited Alexandria, although his permanent home was in Syracuse (in Sicily). Archimedes' interests spanned arithmetic, geometry, optics, statics, hydrostatics, astronomy, and engineering, although his primary interest was geometry.

In geometry, Archimedes proved numerous theorems relating, for example, the surfaces and volumes of spheres and cylinders. (For example, that the surface area of a sphere is four times that of a great circle of the sphere—i.e., $4\pi r^2$.) He also made numerous applications of mathematics to physical problems, such as the determination of the center of gravity of various geometric shapes. He worked on finding areas and volumes of special curved surfaces and solids, by relating them to more easily-found results for triangles, cubes, etc. Some of his procedures involve infinitesimals and anticipate integral calculus. Archimedes' work *On Floating Bodies* presents his famous principle concerning the weight of fluid displaced by an object immersed in the fluid.

As an engineer, Archimedes invented numerous types of geared and levered mechanisms (such as the "Archimedean screw" for lifting water, and the "infinite screw" for translating rotational motion into linear motion), and designed numerous war machines. Two other famous engineers were Ctesibius of Alexandria (fl. c. 270 B.C.) and Hero of Alexandria (fl. c. A.D. 60), who invented water clocks and automatic machines powered by weights, compressed air, heat, and steam.

Another mathematical giant was Apollonius of Perga (fl. 210 B.C.), who wrote a comprehensive and systematic treatise on conic sections, which are the plane figures created when a cone is cut by a plane. He named all three conic sections—the parabola, the ellipse, and the hyperbola—and demonstrated their fundamental properties and methods of construction.

Another subject that intrigued the Greeks was the nature of light and vision. There were vigorous debates about the explanation of how we can see the world around us. Some argued that vision involved something emanating from the eye to external objects. Others (such as the atomists) argued that vision required something coming into the eye, and yet others argued for a combination of the two.

In their studies of optics, the Greeks examined the nature of reflection, refraction, and various types of optical illusion, such as the use of perspective in two-dimensional images. One of the earliest books on optics was written by Euclid, who did experiments with mirrors to demonstrate that the angle of incidence equals the angle of reflection. Studies were also done to see how images were formed on mirrors with curved surfaces, and how curved mirrors could concentrate light from the sun onto focused points.

The astronomer Ptolemy wrote a book on optics which described experiments he performed to measure the angle of refraction when light goes from air into glass, from water into glass, and from air into water. He then investigated the effect this had on the appearance of images (and celestial bodies) when seen through a refractive interface.

In their architecture and artwork, the Greeks showed their aware-ness of various types of optical illusions. The fifth-century scene painter Agatharchus of Samos created flat backdrops that looked three-dimen-sional, with more distant objects painted as smaller. This perspective technique probably did not match the sophistication of the Renaissance, but it was used in numerous relief sculptures, and it was discussed in Euclid's *Optics*, such as in his fifth proposition: "Equal magnitudes situ-ated at different distances from the eye appear unequal, and the nearer always appears larger."[36]

Astronomy

Of all of the scientific achievements of the Greeks, the most significant were in the field of astronomy. The creation of models of the cosmos started with Anaximander in the fifth century. Anaximander's model was extremely primitive, with the earth seen as a cylindrical disk, but soon after this, Greeks such as the Pythagoreans started imagining the earth as a sphere at the center of a spherical cosmos. Greek astronomers promoted the idea of circular motion as the key to the sky, and Greeks worked to explain more complex motions (such as that of planets) in terms of uniform circular motion. The motion of the planets, such as Mars, was especially complex. In the course of moving across the back-ground of the fixed stars, Mars would slow, stop, then move in the oppo-site direction a small distance (a "retrograde" motion), then slow, stop, and then resume its original motion.

The first significant mathematical description of the heavens was given by Eudoxus of Cnidus (a student of Plato) in the fourth century B.C.; Eudoxus created a system of twenty-seven spheres, all of which were centered on the earth. In his system, the earth is at rest in the center, and the axes of the different spheres are inclined to one another. They rotate at different, though uniform, speeds.[37]

With this model, Eudoxus was able to explain the strange motion of the planets. The combination of the motion of several spheres would in principle create the retrograde motion of the planets. The system of

Eudoxus was improved upon by Callippus of Cyzicus, who added seven spheres in order to account for several observational discrepancies.

The systems of Eudoxus and Callippus were primarily mathematical and did not attempt to give a physical explanation of the circular motions. Aristotle took the model of Callippus and made it the basis for a physical system. He was concerned about the transmission of movement from one sphere to the next. He added a number of "reacting" spheres to cancel out the motions of some of the primary spheres, so that each sphere's motion relative to its neighboring spheres would be a simple circular motion. The resulting system had fifty-five spheres.

One notable deficiency of these models was the observed fact that several celestial bodies appear to get bigger or smaller, suggesting that their distance from the earth changes over time. This led Apollonius of Perga to put forth a system using *epicycles*, where the center of each circular motion is itself moving in a circle called its *deferent*.

Once the Greeks had started thinking geometrically about the earth and its relation to the celestial bodies, they were able to apply simple geometrical reasoning to infer certain numerical relations between them. For example, Aristarchus of Samos (c. 310–230 B.C.) measured the relative distances to the sun versus the moon, using a simple calculation. When the moon was precisely half full, and both the sun and moon were in the sky, he measured the angular distance between them. From this angle, he did a simple geometric calculation to determine how much farther the sun is (compared to the distance to the moon). Due to the extreme sensitivity of this calculation on the precision of this angle, the result that Aristarchus calculated was off by an order of magnitude, but his geometric reasoning was flawless. Aristarchus also proposed that the sun is the center of the cosmos, and that the earth orbits it in a circle. While for Aristarchus this was sheer speculation, his idea would later inspire Copernicus to create his own heliocentric hypothesis.

As another example of this geometrical reasoning, Eratosthenes of Cyrene (c. 285–194 B.C.) performed a simple, ingenious calculation to determine the circumference of the earth:

Observing a vertical column at Aswan in Upper Egypt (which lies on the Tropic of Cancer) at noon on the summer solstice, he noted that it cast no shadow. Some 500 miles north in Alexandria at the same moment, the shadow of a similar column formed an angle of seven degrees to the vertical. Eratosthenes reasoned therefore that 500 miles was 7/360, or 1/51, of the circumference of the globe. The result, approximately 25,000 miles is within 6% of the true value.[38]

The most impressive existing artifact of ancient Greek science and engineering is the Antikythera mechanism, which has been called the world's oldest computer. In 1900, a group of sponge divers discovered an ancient shipwreck off the coast of the Aegean island of Antikythera. Dated to approximately 70 B.C., this wreck contained many statues and other artifacts, one of which was a severely corroded mass of bronze apparently containing a geared mechanism with Greek inscriptions. Over the course of the twentieth century, the object was X-rayed and analyzed numerous times; each time, more of the mechanism's features were revealed. The most recent analyses were done in 2006 and 2008 using specialized CT-scan technology.[39]

Researchers have concluded that the Antikythera mechanism was a precisely-constructed astronomical calculator. It contained at least thirty interlocking bronze gearwheels housed in a small wooden box, with knobs that a user could turn by hand. The device was designed to be used by someone who was not an astronomer; it was used to predict solar and lunar eclipses, the phases of the moon, the rising and setting of constellations, and other celestial events.

Analysis of the Greek inscriptions suggested that the mechanism was connected to Syracuse, where Archimedes had lived. Archimedes was known to have produced geared mechanisms, but not of this scale of complexity. Due to the skill involved in constructing this device, it was certainly not the first of its kind. Most probably, some followers of Archimedes had worked on building devices like this, and the Antikythera mechanism was the result of a chain of developments.

In her book on the ancient and modern history of the Antikythera mechanism, author Jo Marchant summarizes what the device could do:

> Whoever turned the handle on the side of its wooden case became master of the cosmos, winding forwards or backwards to see everything about the sky at any chosen moment. Pointers on the front showed the changing positions of the Sun, Moon and planets in the zodiac, the date, as well as the phase of the Moon, while spiral dials on the back showed the month and year according to a combined lunar-solar calendar, and the timing of eclipses. Inscribed text around the front dial revealed which star constellations were rising and setting at each moment, while the writing on the back gave details of the characteristics and location of the predicted eclipses. The mechanism's owner could zoom in on any nearby day—today, tomorrow, last Tuesday—or he could travel far across distant centuries.[40]

The astronomical calculations performed by the Antikythera mechanism made use of cycles that had been identified in part from Babylonian astronomical records. The Greek astronomer who studied these records and created these calculations was Hipparchus of Rhodes, who lived in the first century B.C., and is credited with the transformation of astronomy from a mainly theoretical science into a predictive one.

Hipparchus severely criticized earlier Greek astronomers for the fact that their models did not precisely match astronomical observations. He dedicated himself to integrating the precise Babylonian measurements with the geometrical astronomy of the Greeks. Toward this end, Hipparchus compiled a catalog of 850 stars and their precise celestial longitude and latitude. He discovered the precession of the constellations (the slow shifting in star positions due to a wobble in the Earth's axis).

Hipparchus also invented the dioptra, an instrument for measuring the apparent diameters of the sun and moon. He also created a simple version of trigonometry, which he used to calculate the distance to the moon.

Hipparchus probably also invented stereographic projection, which was an essential element of the astrolabe, and he possibly invented the astrolabe itself. The astrolabe was one of the most critical pre-telescope astronomical instruments, and it was probably invented by Greek astronomers. It consists of a sighting element used to observe the altitude of a celestial body, combined with a representation of the sky to be used for computation.

The pioneering work of Hipparchus provided an essential foundation for the work of the greatest astronomical systematizer of antiquity: Claudius Ptolemy. A Roman citizen of Greek descent, Ptolemy lived in Alexandria in the second century A.D. Ptolemy expanded Hipparchus's star catalog to include 1,022 stars, grouped into forty-eight constellations, with longitude, latitude, and magnitude (from 1 to 6) of each.

Ptolemy's fame stems chiefly from his *Mathematical Composition*, which became known by its Arabic title *Almagest* (from the Arabic word for "the" and the Greek word for "greatest"). *Almagest* is a systematic manual that presented all the astronomy known to the Greeks. It started with simple mathematical and astronomical concepts and built on them step-by-step to explain the types of motions of the celestial bodies, and how mathematical models could account for these motions and phenomena such as eclipses. The celestial bodies included the sun, moon, Mercury, Venus, Mars, Jupiter, Saturn, as well as the fixed stars, which were believed to lie on one sphere centered on the earth. For each type of motion, Ptolemy proposed a geometric model, calculated the numerical parameters from observations, and then constructed tables enabling the prediction of specific positions and phenomena for a given date.

Ptolemy's geometric models made use of the epicycles developed by Apollonius and added eccentric motions operating in conjunction with them.[41] An eccentric motion involves an off-center point around which there is uniform angular motion, which leads to a non-uniform circular motion. While this last element would be criticized by later astronomers (because of its departure from uniform circular motion), it was necessary to achieve the high level of precision and predictive success of Ptolemy's system. And according to one historian of science, "as a didactic work the Almagest is a masterpiece of clarity and method, superior to any ancient scientific textbook and with few peers from any period."[42] Because of its tremendous and obvious value, it was the most influential work of astronomy until the *De Revolutionibus* of Copernicus was published in 1543.

* * *

As we have seen, Ancient Greeks made many impressive advances in a range of different scientific fields. While many of their theories are deficient by the standards of today's knowledge, we must remember that they started with a pre-scientific culture. From there, they got science started. They did perform experiments in a number of fields (music, optics, mechanics, physiology), and they did use mathematics in a number of fields (music, optics, mechanics, astronomy). Of course, they did not systematize the use of mathematics and experimentation to the extent that would come in the seventeenth century with the Scientific Revolution.

The Ancient Greeks created science, and they could not have done so within a fundamentally religious culture. Essential for the creation of science was a specific view of the nature of reality—a specific metaphysics: As physicist David Harriman points out:

> The impersonal metaphysics was the great — and historically recent — achievement of the Greeks, specifically of Aristotle's secularism and advocacy of reason. It was this approach that led to the clear Greek distinction between the animate and the inanimate, which included the fact that consciousness can belong only to the animate. Once the Greek approach was embedded in the mind of the West . . . causation could no longer be conceived in terms of the personal efficacy of supernatural agents. Thus did the objective view of cause and effect displace the anthropomorphic view that, at the beginning, had seemed to be merely an innocent extension to nature of men's own causal experience.

Western civilization broadened the concept of "cause," by regarding personal efficacy as merely a subtype of it. This was a crucial precondition of the development of modern science. It amounts to bringing causality for the first time into the realm of reality and identity — i.e., to breaking its primordial bonds to mysticism.[43]

CHAPTER THREE

The Disappearance of Science in the Age of Faith

As WE SAW in the last chapter, scientific work continued in the six centuries after Aristotle in Alexandria and elsewhere, and even reached new heights in figures such as Ptolemy and Galen. But during this same period, an ominous trend was underway: the main currents of classical philosophical thought were gradually turning *against* reason and science.

The Decline of Greek Philosophy after Aristotle

AFTER ARISTOTLE'S DEATH, both Plato's Academy and Aristotle's Lyceum continued to function and to teach the philosophic systems of Plato and Aristotle, respectively. But three new schools of Greek philosophy became dominant: Epicureanism, Stoicism, and Skepticism. Their ideas were relatively unoriginal and heavily influenced by Plato.

The political situation greatly influenced Greek philosophic thought at this time. For many centuries in Greece, the independent city-state had been the natural center of the political world. With the conquest of Alexander and then the Romans, however, it looked like the era of the city-state was over. Numerous traumatic wars and social upheavals accompanied this change, and the resulting anxiety was reflected in the philosophic thought of the time.

The school of Epicureanism (founded in 306 B.C.) was named after a teacher—Epicurus—who believed that the best life is a simple one. His teachings continued the atomist tradition of Democritus (c. 460–370 B.C.) who held that everything consists of atoms and the void. The ideas of Epicurus were later popularized in the Roman world by the poet and philosopher Lucretius (c. 99–55 B.C.) in his book *On the Nature of Things.*

Epicurus's psychological and ethical outlook was the most significant and original aspect of his philosophy. Epicurus stated that the goal of life is to maximize pleasure and avoid pain. However, he believed that pain and suffering are prevalent in the world, and as a result, he advocated for a life of withdrawal. Actively striving for goals is too risky, he held, and is bound to lead to failure and heartbreak. (Epicurus was far from an *epicure* in the modern sense.)

Historian of philosophy W. T. Jones suggests that "despite what Epicurus said, the supreme good in his system was not pleasure as such, but repose. . . . An ethics that emphasized quiet and repose appealed to men who had abandoned the ideal of all-round personal development because they had found the society they lived in too complex for them to hope to control."[44]

The philosophy of Stoicism was founded by Zeno of Citium (c. 335–264 B.C.), who taught that the universe has a soul—the World Soul—who is everywhere and in everything, including us. This view, known as pantheism, led to the idea that we are not autonomous creatures, separate individuals, since we are a part of the World Soul.

In ethics, the Stoics were much more consistent than Epicurus in their asceticism and self-denial. Stoics held that we should stop valuing pleasure, friends, and even life. We must achieve *apathy*—the peace of mind that comes through the acceptance of the universe as it is, including an utter indifference to the course of events. But at the same time, the stoics placed great importance on the concept of *duty*: we have duties that we must obey, simply because they are our duties.

Famous Roman Stoics included the orator Cicero, the slave Epictetus, and the emperor Marcus Aurelius. The fact that this philosophy was promulgated by both a Roman slave and a Roman emperor demonstrates its dominance in the culture.

The Skeptical school of philosophy, started by Pyrrho of Elis (c. 360–270 B.C.), was much less organized than the Epicureans' and the Stoics'.[45] Descended from the sophists against whom Plato had argued, the skeptics claimed that we can know nothing (and, they added, we can't even know *that*). One of their arguments relied on the relativity of perception. Suppose two buckets of water sit in front of you, one with cold water and the other with hot. Suppose you soak one hand in the cold water while soaking the other hand in the hot water. You then take both hands out and place them both in a bucket of lukewarm water. The hand that was originally in the hot water will now feel cold, and the other will feel hot. So, is the third bucket of water cold or hot? How can we ever know for sure if our senses are deceiving us?

The more consistent of these thinkers applied their skepticism to their ethical views: we cannot know, they concluded, what is right and what is wrong, so there is no point in worrying about doing the "right" thing. We should simply do what is convenient. There is no point in pursuing goals since we cannot know if we will achieve them.

So what was the significance of these three schools? Leonard Peikoff answered this in his lectures on the history of philosophy:

> All these schools (the Epicureans, Stoics, Skeptics) are converging on the same conclusion: The helplessness of man, and the hopelessness of life. Each of these schools in its own way is ripening man for the onset of a religious era. Each is sapping man's confidence in some vital area, and preparing him to fall to his knees and seek divine guidance, divine knowledge, other-worldly happiness. . . .
>
> We see philosophy progressively losing confidence; man's mind is no longer regarded as capable of gaining knowledge; the senses are invalid; reason is precarious, unreliable; life on earth is hell, inherently painful, malevolent. We have to give up the hope of happiness on earth.[46]

There was one more significant pagan philosopher before the complete eclipse of Greek secular philosophy, but he was in many respects more of a mystic than a philosopher.

The Mystical Philosophy of Plotinus

PLATO HAD PROVIDED a philosophic foundation for mysticism, but Plato's mysticism would be taken one step further by a philosopher named Plotinus. Originally from Egypt, Plotinus (c. A.D. 204–270) settled in Rome and taught a philosophy which became known as Neoplatonism, and which would become a powerful influence on the fathers of Christianity.

Plotinus fully accepted Plato's arguments for the existence of another world—a superior dimension—beyond the world we see. But Plotinus held that there is not just one such dimension; there are three of them. Historian Edward Grant gives a good summary of this philosophy:

> For Plotinus, God, a transcendent being called the One, is beyond all being of which we can have any experience. God is beyond all distinctions. Moreover, God does not engage in thought or in any acts of the will. No positive attributes can be assigned to the One because to do so would be to delimit and particularize God. God is absolutely unchangeable and omnipresent. And yet the world emanates from him by necessity. . . . The first emanation from God is Mind or Thought . . . which is eternal and beyond time. From Mind or Thought emanates the Soul, akin to Plato's World Soul, and from Soul emanates individual souls. . . . In the eyes of Plotinus, the material world has no positive qualities; indeed it is utterly negative.[47]

In order to behold the One (which transcends anything and everything we can experience), you must undergo a long, elaborate process. According to Plotinus, you have to, in effect, jump outside yourself and the entire physical world. This state—mystical union with the One—came to be called "ecstasy" ("standing outside").

The Mystery Cults of the Roman Empire

THE ROMANS WERE heavily influenced by Greek thought. But the more abstract of the Greek ideas were not of much interest to them; the

Romans were not thinkers, but doers. No significant science was done by a Roman citizen who was not culturally Greek.

There was no significant Roman philosopher who defended Rome while upholding reason and this world. The Romans fell under the spell of Stoicism, with its emphasis on duty, asceticism, and otherworldliness, and as time went on, Stoicism became even more obsessed with the next world.

The Roman conquest of the lands around the Mediterranean led to cultural intermixing, which led to a variety of new religions, as gods and goddesses from one culture spread to wider areas. In the late classical world, large numbers of people became interested in so-called "mystery cults," which offered their worshippers a path to salvation and eternal life. They were called "mystery" cults because they typically performed ceremonies open only to those who had been initiated into their secret rituals. All of these religions emphasized our sinful nature and the necessity for purification; all of them preached various forms of self-denial; each promised a glorious immortality to its dedicated followers.

There was a large number of these cults, but three were particularly important: The Great Mother of Phrygia, the cult of Isis and Osiris, and Mithraism. The cult of the Great Mother of Phrygia was based on the myth of the goddess Cybele:

> When her lover, Attis, died, the goddess mourned, and death came upon the world. When Attis was brought back to life, the goddess rejoiced and nature put on a garment of green. To rehearse the myth was doubtless at first simply a way to assure a good crop. Later, when it became a way for the worshippers to share in the immortality of Attis, the old vegetation rites were retained. In the spring of the year the adherents of the cult indulged in a period of mourning for the dead Attis. They fasted and flagellated themselves; more passionate devotees castrated themselves in a frenzy of excitement. These latter worshippers, exalted by their sacrifice, became priests of the cult.[48]

The cult of Isis and Osiris, originally from Egypt, was centered on a similar myth, in which the god Osiris dies and is later resurrected.

Mithraism, originally from Persia, had a more complex theology, in which the world is seen as a battleground between two supernatural forces, which are good and evil powers. The god Mithra stands on the side of the good and rewards his supporters after they die.

By the third century A.D., belief in the Roman gods had withered, and people were looking for substitutes. As a result, the mystery cults grew rapidly. But they had serious competition from a relatively new religion that had its origins in a sect of the ancient monotheistic religion of Judaism.

The Birth of Christianity and the Idea of Faith

THE FOUNDER OF this sect was the Jewish preacher Jesus (4 B.C.–A.D. 29). Jesus saw himself as a reformer of Judaism, and he advocated turning away from overly legalistic interpretations of the Jewish laws. Jesus strongly believed that the end of the world was near—within the lifetimes of those listening to him. So, he urged: "Repent ye, for the Kingdom of God is at hand." Don't prepare for tomorrow, because there will be no tomorrow. Prepare your soul for God's judgment. Obey the spirit of the Law, and practice the virtues of humility, forgiveness, charity, faith, and the willing acceptance of Yahweh's commands.

This morality was in striking contrast to the morality of ancient Greece as embodied in Aristotle's philosophy. Aristotle had argued that the good life consisted in pursuing excellence, in realizing the full potentialities of human nature, in rationality, in material success, and in a justly-deserved pride. For Aristotle, pride was the crown of the virtues, but for Jesus, it was a terrible sin. Moreover, as W. T. Jones points out, "There is nothing in Jesus' list of virtues that corresponds to Aristotle's 'intellectual virtues' —science, art, philosophic wisdom, and so on."[49]

The religion known as Christianity developed gradually, out of the teachings of Jesus combined with elements from Judaism, the other mystery cults, and late Greek philosophy. One of the most influential figures in this development was the apostle Paul (d. c. A.D. 62). Paul transformed the Jesus cult into the religion of Christianity—a religion focused on mystical union with Christ. Paul taught that God created the first man, Adam, free from sin. But Adam disobeyed God, and we have inherited Adam's sin. As the sin of one man (Adam) brought suffering and death into the world, so the virtue of one man (Jesus) saves us. We need to embrace Jesus as our lord and savior in order to save our eternal souls.

According to Paul, a major distraction that can prevent us from embracing Jesus is an excessive attachment to the pleasures and wonders of this world: "The wisdom of this world is foolishness with God."[50] So he advises: "Be on your guard; do not let your minds be captured by hollow and delusive speculations, based on traditions of man-made teaching centered on the elements of the natural world and not on Christ."[51]

From Paul onwards, faith was considered a central part of Christianity, and many Christians openly rejected rational thought and empirical evidence. They often happily pointed out their own lack of education, associating independent reasoning with the sin of pride. According to historian Charles Freeman:

> The effects of Paul's condemnation of "the philosophers" could not have been put more clearly than by John Chrysostom, an enthusiastic follower of Paul. "Restrain our own reasoning, and empty our mind of secular learning, in order to provide a mind swept clear for the reception of divine words." [Bishop] Basil echoes him: "Let us Christians prefer the simplicity of our faith to the demonstrations of human reason. . . . For to spend much time on research about the essence of things would not serve the edification of the church." This represented no less than a total abdication of independent intellectual thought, and it resulted in a turning away from any speculation about the natural world. . . . "What purpose does knowledge serve – for as to knowledge of natural causes, what blessing is there for me if I should know

where the Nile rises, or whatever under the heavens the 'scientists' rave about?" wrote Lactantius in the early fourth century. One Philastrius of Brescia implicitly declared that the search for empirical knowledge was itself a heresy:

"There is a certain heresy concerning earthquakes that they come not from God's command, but, it is thought, from the very nature of the elements. . . . Paying no attention to God's power, they [the heretics] presume to attribute the motions of force to the elements of nature . . . like certain foolish philosophers who, ascribing this to nature, know not the power of God."[52]

Christian doctrine took shape gradually, as numerous incompatible Gospels competed for adherents. There was a wide range of beliefs and practices among the early Christian communities, but a set of core beliefs emerged, including: (1) the affirmation of God the Father and creator of the universe; (2) Jesus the Son of God, whose death and resurrection had raised the possibility of salvation for all who repented, and who is God; (3) a Holy Spirit who continues to act as a divine force in the world, and is also God.

Given the inherent arbitrariness of these ideas—their disconnect from logic and evidence—questions arose almost immediately about their meaning. What was the relationship between the Son and the Father? If Jesus was God, then how could he suffer on the cross? How could God be both three and one? Different sects accepted many different answers to these questions, and most of these answers would eventually be considered heretical.

The Patristic Fathers of Christianity

Tertullian

One of the most influential of the early Christian theologians came from a pagan family in Carthage, in northern Africa. Tertullian (c. A.D. 150–c. 225) received a thorough education in classical philosophy, then converted to Christianity and became a strident critic of his earlier learning and a highly influential defender of orthodox Christian doctrine. He excoriated philosophy as the incubator of heresies and lashed out at "wretched Aristotle" and especially his writings on logical arguments.

In a famous passage, Tertullian asks:

> What indeed has Athens to do with Jerusalem? What concord is there between the Academy and the Church? What between heretics and Christians? Our instruction comes from "the porch of Solomon," who taught that "the Lord should be sought in simplicity of heart." Away with all attempts to produce a mottled Christianity of Stoic, Platonic, and dialectic composition! We want no curious disputation after possessing Christ Jesus, no inquisition after enjoying the gospel! With our faith, we desire no further belief. For once we believe this, there is nothing else that we ought to believe.[53]

Showing his disdain for logic and reason, Tertullian wrote that we must believe Christian claims about the resurrection precisely because they are senseless: "The Son of God was crucified; I am not ashamed because men must needs be ashamed of it. And the Son of God died; it is by all means to be believed, because it is absurd. And He was buried, and rose again; the fact is certain, because it is impossible."[54]

Ambrose

Another influential theologian of early Christianity was Saint Ambrose, Bishop of Milan (c. A.D. 340–397). Ambrose clearly stated his disapproval of the Greek astronomers and their attempts to understand the structure of the cosmos:

> To discuss the nature and position of the earth does not help us in our hope of the life to come. It is enough to know what Scripture states, "that He hung up the earth upon nothing" (Job 26.7). Why then argue whether He hung it up in air or upon the water [the views of Anaximenes and Thales]. . . . Not because the earth is in the middle, as if suspended on even balance [the position of Anaximander], but because the majesty of God constrains it by the law of His will, does it endure stable upon the unstable and

the void. The Prophet David also bears witness to this when he says: "He has founded the earth upon its own bases: it shall not be moved for ever and ever." (Ps. 103.5)[55]

Showing the influence of the Neoplatonist conception of the "One," Ambrose stated that God is ultimately "a divine power incomprehensible to human minds and incapable of being expressed in our language."[56] Ambrose considered natural philosophers to be fools: "'Man is become a fool for knowledge.' (Jer. 10.11–14) How can one who pursues the corruptible things of the world and thinks that from these things he can comprehend the truth of divine nature not become a fool as he makes use of the artifices of sophistry?"[57]

Augustine

Perhaps the most influential theologian of Christianity has been Saint Augustine (A.D. 354–430). Augustine converted to Christianity in 386 and was baptized the following year in Milan by Saint Ambrose. Augustine became bishop of Hippo, and he worked to integrate Christianity with the philosophic framework of Neoplatonism. Augustine's synthesis of Plato with Christianity gave the religion a theoretical framework, and Plato's ideas reinforced the new authoritarianism. The Platonists held that only a few could grasp the reality of the immaterial world, including the true nature of the Form of the Good; these privileged few could prescribe the good for everyone else. Augustine used this idea to support the idea of Church authority.

Augustine saw the events of the world as, in effect, a grand play in which God is the playwright, the producer, the set designer, and the director, and we are the actors. As the controller of the events of the play, God has introduced symbols that we may be able to discover; it is mainly in this sense that Augustine approved of learning from the events of history and the facts of this world.

When it came to scientific matters like what lies beyond Saturn, he held that "The authority of Scripture is greater" than "all human ingenuity."[58] Augustine was not interested in the physical world; he possessed none of the natural curiosity of the Greeks. Instead, he had a fear of falling into intellectual pride for prying into God's secrets:

> When it is asked what we ought to believe in matters of religion, the answer is not to be sought in the exploration of the nature of things, after the manner of those whom the Greeks called "physicists." Nor should we be dismayed if Christians are ignorant about the properties and the number of the basic elements of nature, or about the motion, order, and deviations of the stars, the map of the heavens, the kinds and nature of animals, plants, stones, springs, rivers, and mountains; about the divisions of space and time, about the signs of impending storms, and the myriad other things which these "physicists" have come to understand, or think they have. . . . For the Christian, it is enough to believe that the cause of all created things, whether in heaven or on earth, whether visible or invisible, is nothing other than the goodness of the Creator, who is the one and true God.[59]

Historians have sometimes argued that Augustine was not categorically opposed to studying the natural world. David Lindberg argues that "there were contexts in which Augustine's attitude toward pagan works on natural philosophy was relatively favorable."[60] Lindberg quotes from Augustine:

> Even non-Christians know something about the earth, the heavens, and the other elements of this world, about the motion and orbit of the stars and even their size and relative positions, about the predictable eclipses of the sun and moon, the cycles of the years and the seasons, about the kinds of animals, shrubs, stones, and so forth . . . Now, it is a disgraceful and dangerous thing for an infidel to hear a Christian, presumably giving the meaning of Holy Scripture, talking nonsense on these topics; and we should take all means to prevent such an embarrassing situation, in which people show up vast ignorance in a Christian and laugh it to scorn.[61]

In addition to the value of helping Christians avoid embarrassment in front of infidels, natural knowledge, Augustine held, could be useful for helping us interpret certain Biblical passages, and for calculating the correct date of Easter. But overall, Augustine saw natural knowledge as a handmaiden for religion, and as having no value apart from religious purposes. Late in life, he regretted ever emphasizing the study of the seven liberal arts, and he argued that natural knowledge is of no value to a Christian.

Like earlier Church fathers, Augustine attacked the intellectual self-confidence of the natural philosophers, who

> lapse into pride without respect for you, my God, and fall into shadow away from your light, but although they can predict an eclipse of the sun so far ahead, they cannot see that they themselves are already in the shadow of eclipse. This is because they ignore you. . . . Their conceit soars like a bird; their curiosity probes the deepest secrets of nature like a fish that swims in the sea; and their lust grows fat like a beast at pasture. . . . Oh, God, you are the consuming fire that can burn away their love for these things and re-create the men in immortal life.[62]

For Augustine, scientific curiosity about the natural world was merely another variety of sensual temptation:

> The mind is also subject to a certain propensity to use the sense of the body, not for self-indulgence of a physical kind, but for the satisfaction of its own inquisitiveness. This futile curiosity masquerades under the name of science and learning, and since it derives from our thirst for knowledge and sight is the principle sense by which knowledge is acquired, in the Scriptures it is called *gratification of the eye*. (1 John 2:16)[63]

In 529 at the dedication of a church in Gaul, the Council of Orange, attended by fourteen bishops, affirmed the theology of Augustine, and other councils followed suit. Augustine would become the single most important theologian of Christianity, eclipsed by Aquinas only centuries later during the Scholastic period.

The Murder of the Remnants of Pagan Philosophy

PAGAN GREEK PHILOSOPHY had gradually become more mystical, more skeptical about knowledge of this world, and more ascetic. In short, it had turned against reality, reason, and human life. As Christianity had grown, especially in the second century, intellectually-minded individuals had tended to turn away from pagan philosophy and toward Christian apologetics. But some interest in pagan philosophy—especially Neoplatonism—had continued, and schools continued to teach it.

The final death of pagan philosophy was the result of an active campaign of suppression on the part of Christian authorities. Rome played a key role in this suppression, and the guiltiest Roman emperor was Constantine. This story is told dramatically by Charles Freeman in his book *The Closing of the Western Mind: The Rise of Faith and the Fall of Reason.*

Before the rise of Christianity, Rome had been relatively tolerant of different religious and philosophic traditions (Judaism being the major exception). As long as people acknowledged a basic level of respect for the official gods of Rome, they were free to practice their own religion and to worship their own gods in virtually any way they pleased. Persecutions of the Christians were limited in time and extent, and were mainly confined to the reign of Diocletian (Emperor 299–311).

It was Constantine who began the Christianization of the Roman State. Seeing the popular religion as a unifying force, he decided to adopt Christianity for political reasons. Constantine's Edict of Milan (in 313) ensured that Christianity would be tolerated throughout the Roman Empire, and in itself, this edict was a positive step for religious freedom. However, he did not stop with this edict but used the Roman state to actively promote Christianity, starting with the granting of special legal favors to Christian clergy.

But the popular religion was not as unifying as Constantine thought it would be. He soon discovered that there were many different communities, with different beliefs, who all called themselves Christian. These differences led to numerous theological disputes. The sometimes-bitter controversies threatened the political stability for which Constantine had adopted Christianity in the first place. A particularly bitter debate concerned the bishops who were influenced by the teachings of Arius, who held that Jesus was a subordinate divinity distinct from God the Father. This view directly conflicted with that of Athanasius, the bishop of Alexandria, who held that Jesus was in no way subordinate. The controversy spread as other bishops associated with one side or the other. To resolve this issue, Constantine called a council of bishops—the Council of Nicaea—in order to decide on the correct Christian doctrine, to be supported by the state.

In 381, the emperor Theodosius decreed the doctrine of the Trinity to be orthodox and expelled dissenters from their churches. At the Council of Constantinople of the same year, he issued a new edict defining the tenets of Christianity and the consequences for those who disagreed:

> We authorise the followers of this law to assume the title of orthodox Christians; but as for the others, since in our judgment, they are foolish madmen, we decree that they shall be branded with the ignominious names of heretics, and shall not presume to give to their conventicles the names of churches. They will suffer in the first place the chastisement of divine condemnation, and in the second the punishment which our authority, in accordance with the will of heaven, shall decide to inflict.[64]

Enormous amounts of state financial support became available for the building and maintenance of approved churches. Bishops were tied into the legal system and given a wide range of social roles, including "spiritual leader, patron, estates manager, builder, overseer of law and order, city representative, and protector of the poor."[65] After worship at pagan shrines was, in effect, banned, Christian mobs went to work destroying the ancient shrines.

It was finally under Justinian (Emperor 527–65) that the law was fully directed against dissenters from Christian doctrine. The death penalty was the official punishment for participation in pagan cults. Pagan teachers (including philosophers) were forbidden by law from further teaching. Plato's Academy, after 900 years of teaching, was finally closed in 529.[66]

The Greek tradition had officially been outlawed, and it was the beginning of the Dark Ages, an era dominated by "incessant warfare, corruption, lawlessness, obsession with strange myths, and an almost impenetrable mindlessness."[67] Belief in bizarre and miraculous events became widespread, even among the most educated. The intellectual self-confidence and curiosity of the Ancient Greek world had been extinguished and labeled as the dreaded sin of pride. Faith and obedience to the authority of the Church were now more important than reasoned thought. To the extent that there now was any intellectual activity, it was a rationalistic, deductive "logic" that was utterly divorced from reality. Science was virtually absent from the West for the next millennium.

CHAPTER FOUR

Reason as the Existence-Oriented Faculty

As we saw in Chapter One, central to all science is the use of *reason*—"the faculty that identifies and integrates the material provided by man's senses."[68] The validity of science relies on the validity of reason.

A full validation of reason lies beyond the scope of this book.[69] But it is necessary here to identify several fundamental ideas that are at the base of reason and all of our knowledge, even if they are rarely identified explicitly. The following sections form the most abstract part of this book, but they are critical for illuminating the chapters that follow.

Are any Truths Self-Evident?

The idea of self-evident truths is not popular among today's intellectuals. In the past, many thinkers have claimed to find "self-evident" truths that were anything but. Descartes claimed any idea that was "clear and distinct" to be, by that fact alone, true. Many theists have argued that the existence of God is self-evident. Many thinkers have claimed that certain ethical ideas, such as the goodness of altruism, are self-evident. However, none of the above ideas comes close to qualifying as truly self-evident.

If there are no self-evident truths to serve as starting points for our knowledge (as most modern philosophers hold), then our knowledge becomes a structure built on quicksand, and any hope of finding certainty on any issue is lost. Any hope we have of defending science is also lost, and, like philosopher Paul Feyerabend, we will have no basis on which to prefer modern science over voodoo.[70]

Axioms of Philosophy

IN LIGHT OF the critical need to ground our knowledge on a sound foundation, I ask the reader to set aside any pre-existing aversion he may have toward the idea of "self-evident truths" and to carefully consider the following presentation of axioms as identified by philosopher and author Ayn Rand.

At the base of philosophy and all our knowledge are three ideas that are so basic and irreducible, that Rand called them axioms. The first of these is the concept of "existence:"

> The first point to be made is: *Existence exists.* A rational philosophy starts with the most fundamental point. Looking out at the world, at any aspect of it—in this case, at a computer—we say: this *is*. Looking out the window, at the trees, buildings, and automobiles in the distance, we say: these things *are*. Something exists.

The fact of existence is not only self-evident; it is an irreducible starting point for all cognition:

> Prior to any subsequent question, before one can even seek to identify what kinds of things there are—before any further learning takes place, there must first be something to be learned and one must know this basic fact. "If not, there is nothing to consider or to know."[71]

Note that this is not the same thing as the idea that a physical world exists. The concept "existence" does not indicate what kind of things exist; it merely stresses the primary fact that they *exist*.

The next axiom is the fact that we have the faculty of consciousness:

> Existence exists—and the act of grasping that statement implies two corollary axioms: that something exists which one perceives and that one exists possessing consciousness, consciousness being the faculty of perceiving that which exists.[72]

Of course, a world without consciousness is possible. But in order for you to be aware of that world, you must be conscious. In fact, "aware-ness" is simply another word for "consciousness."[73]

The third fundamental axiom is the *law of identity*: "This principle states that a thing is what it is—and is not what it is not. Symbolically: A is A—and it is not non-A. For any existent, to be is to be *something*."[74] As Rand points out, "The concept 'identity' does not indicate the par-ticular natures of the existents it subsumes: it merely underscores the primary fact that *they are what they are*."[75] Instead of saying "existence has identity," it is more accurate to say "existence *is* identity."[76] To be is to be something specific.

An important idea that follows directly from the above axioms is the *law of causality*, which states that the identity of an entity makes it act in a specific way. An entity's *identity* is the *cause* of its *actions*. The order, lawfulness, and regularity we see in the world comes from the fact of causality, which is inherent in reality as such. To be is to be *something*, and thereby to act accordingly.

The Primacy of Existence versus
the Primacy of Consciousness

THE AXIOMS OF existence and consciousness have a natural order: Existence comes first, then consciousness. Rand called this fact the *pri-macy of existence*:

> Existence, this principle declares, comes first. Things are what they are independent of consciousness—of anyone's perceptions, images, ideas, feelings. Consciousness, by contrast, is a depen-dent. Its function is not to create or control existence, but to be a spectator: to look out, to perceive, to grasp that which is.[77]

The opposite of this principle is the *primacy of consciousness*: the idea that consciousness creates reality. Imagine a primitive farmer who des-perately wishes for rain for his crops. He stands in his field, turns toward the sky, closes his eyes, starts chanting magic incantations, and imag-ines rain falling. He is operating on the premise that his consciousness controls existence. The primacy-of-existence mentality, on the other hand, would recognize that wishes don't directly control reality, that the weather is what it is, and that there may be other relevant facts that he can identify, perhaps about other sources of water and how he can make use of them.

The *primacy of consciousness* comes in three main variants, depending on whose consciousness is seen as creating reality. There is the personal form, in which one's own mind creates reality; there is the divine form, in which God creates reality; and finally, there is the social form, in which society's collective mind (or a majority vote) creates reality.

Looking back at the last two chapters, we can detect these philosophic views among various thinkers in Ancient Greece and the Dark Ages.

Plato's philosophy was based on the *primacy of consciousness*. Plato's Forms were in essence mental objects—aspects of consciousness. The Neoplatonists, and later Augustine, held that the Forms are ideas in the mind of God. Moreover, per Plotinus, the world we live in is the lowest level of being; it is merely an emanation from mystical consciousnesses—the World Soul, the World Mind, and God.

In contrast, Aristotle's philosophy was implicitly based on the *primacy of existence*. For Aristotle, reality was not Plato's mental/mystical Forms, but the individual objects we see around us. For Aristotle, these things exist; they are what they are, independent of anyone's awareness of them.

The Ancient Greeks created science, and they could not have done so within a fundamentally religious culture. Essential for the creation of science was a particular view of the nature of reality—a certain metaphysics—that accepts the *primacy of existence*.

Christianity, from its very beginning, with its God-centered view of the world, had the *primacy of consciousness*—of God's consciousness—as its philosophic fundamental.

Are These Axioms Falsifiable?

IN THE ABOVE sections, readers familiar with the ideas of philosopher Karl Popper may be prompted to ask, "But are these axioms falsifiable?" Popper originated the idea of falsifiability as a way of demarcating between scientific claims and pseudo-scientific (or meaningless) claims.

A falsifiable claim, according to Popper, is one that can be put to the test of observation, and for which it is conceivable that a disproving observation could be made. For example, the claim that "all swans are white" would be disproven by the observation of a black swan. Popper held that good science involves the creation of broad falsifiable claims followed by systematic attempts to falsify them. In proposing this idea, he was motivated by the reasonable concern to exclude from science those hypotheses that could never be tested.

The axioms of existence, consciousness, and identity are clearly not falsifiable, since there are no conceivable observations which could prove them wrong. So, are these ideas pseudo-scientific or meaningless? Not at all.

As we saw, these philosophic axioms are preconditions of science and all rational thought. Denying any of them leads to an absurdity. Suppose that someone tries to deny the axiom of existence: He says, "nothing exists." Then by the logic of his own statement, he doesn't exist, and his statement doesn't exist.

Suppose he tries to deny that consciousness exists. He says, "nobody is conscious." Then by the logic of his own statement, he has emitted a meaningless sequence of sounds to an audience that is not aware of them.

Suppose he tries to deny the axiom of identity. He says, "things are not what they are." Then his statement is not what it is; it is not even a denial.

Clearly, it does not make sense to deny these axioms.[78] But does it then make sense to affirm them?

Yes, it does, especially when one is tempted to implicitly reject them. Consider a situation in which one has become aware of some painful facts: A woman discovers that her husband is having an affair. She can try to evade what she knows by excessive drinking or practicing other forms of denial, or she can say to herself, "this is a fact, I know what I know, it is what it is;" and then take some action to deal with the situation effectively.

Science and Religion as Fundamentally Opposite

In Chapter One, we looked into the philosophic roots of science and religion to see that they are fundamentally in conflict. The clash is clear in both metaphysics and epistemology. In metaphysics, the clash is between the otherworldly view of religion and the this-worldly view of science. In epistemology, the clash is between faith and reason:

> [Faith] is the essential that distinguishes religion from science. A scientist may believe in entities which he cannot observe, such as atoms or electrons, but he can do so only if he proves their existence logically, by inference from the things he does observe. A religious man, however, believes in "some higher unseen power" which he cannot observe and cannot logically prove.[79]

But we can now go to a deeper level of explanation than this. At the bottom is a clash about the axiomatic base of philosophy. *The validity of faith relies on the primacy of consciousness; the validity of reason rests on the primacy of existence. This is the deepest root of the clash between science and religion.*

In the following chapters, as we examine more episodes in the history of science, we will see this philosophical clash continue to manifest itself, albeit sometimes in non-obvious ways.

CHAPTER FIVE

The Rise and Fall of Science in the Islamic World

As Europe descended into its Dark Ages, and its intellectual activity was reduced to monks copying books that they barely understood, the rise of a new religion had momentous implications for the history of science.

The Translation Movement

THE RELIGION FOUNDED by Muhammad was named "Islam," which means "submission"—to the will of God (Allah). It stresses the *unity* of God—as opposed to the Christian trinity—and the Quran as the uncreated word of God, existing in heaven in Arabic. Muhammad died in A.D. 632, and within one century Muslims had violently conquered large parts of the Middle East, North Africa, the Iberian Peninsula, and parts of India, and had united them into a single empire, under a single religion and language.

As the Islamic world expanded into the Middle East, it came into contact with the scientific achievements of the Greeks. Even before the rise of Islam, Arabic astronomers in various Middle Eastern centers had learned about Greek astronomy. When Islam came, Muslims immediately saw astronomy as a science to be used in the service of Islam; it provided techniques for determining the times of prayer, the beginning and end of Ramadan, and the *qibla* (the direction towards Mecca for prayer).

A massive translation movement, lasting over two centuries, was given initial support by the Abbāsid dynasty's early caliphs. Of these, it was the caliph al-Ma'mun (reign 813–833) who provided the greatest support for the translations and resulting intellectual activity. Following a dream in which Aristotle appeared to him, al-Ma'mun created an institution for translation in Baghdad in 830, which became known as the House of Wisdom. This became a great center of translations from Greek as well as Persian and Sanskrit. Al-Ma'mun was an enthusiastic supporter of rational discussion of the newly discovered ideas, and he even hosted debates in his court between Christians and Muslims. Al-Ma'mun also supported al-Kindi—considered the first major philosopher of Islam—who worked to synthesize the ideas of Plato with Islam.

One new technology greatly helped the translation movement. When the Abbāsid army defeated the Chinese in 751 near Samarkand, their Chinese prisoners of war included those who were familiar with papermaking. (Paper was a Chinese invention from the second century A.D.) Soon, paper mills were being built in Baghdad. Paper was much cheaper than papyrus and parchment, so it was widely used for the new translations.

The translation movement was funded by not only the caliph; many wealthy patrons paid out large amounts of money for translations. For those who could do it, translation became a lucrative profession. The first translations were of works containing the most obvious practical value: astronomy and medicine. These were followed by other scientific topics, and then philosophy. (The main subjects that were not translated were history, poetry, and drama.) By 1000, virtually all of the writings of Greek mathematics, medicine, and natural philosophy had been translated into Arabic.

To their credit, numerous Muslims became captivated by the Greek world:

> Here . . . a new world appeared: one in which men had reasoned
> fearlessly about everything, unchecked by sacred scriptures, and
> had conceived a cosmos not of divine whimsy and incalculable
> miracle, but of majestic and omnipresent law. . . . [N]ow for three

centuries Islam played the new game of logic, drunk . . . with the "dear delight" of philosophy.[80]

The Scientific Achievements of the Islamic World

THE SCIENTIFIC ACHIEVEMENTS of the Islamic world fall into four categories: mathematics, astronomy, optics, and medicine. In each of these areas, thinkers worked first to master the new ideas, then went on to make improvements and new discoveries.

The most accomplished mathematician of the Islamic world was al-Khwārizmi (c. 780–c. 850), a Persian who served at the court of the caliph al-Ma'mun in Baghdad, and who is known as the father of algebra. Greek mathematicians (such as the third-century Alexandrian Diophantus) had worked out solutions for certain algebra-like word problems, even using symbols for the unknown quantities. But al-Khwārizmi took things a step further:

> What al-Khwārizmi did for the very first time and what sets him apart from all other mathematicians before him is subtle but crucial. He abandons the practice of solving particular problems, and instead provides a general series of principles and rules for dealing with (quadratic) equations, solving them in a series of steps: the algorithm. . . . Although al-Khwārizmi uses words rather than symbols as Diophantus did, he is in fact much closer to the sort of algebra we do today than the Greek is, because, for al-Khwarizmi, the unknown quantity . . . is a new kind of object which can be manipulated in its own right.[81]

The word "algorithm" comes from the Latinized version of al-Khwārizmi's name, and the word "algebra" comes from the Arabic term "al-jebr" (meaning "restoration" or "completion") from the title of his book on the subject: *al-Kitab al-Mukhtasar fi Hisāb al-Jebr wal-Muqābala* ("The Compendium on Calculation by Restoration and Balancing"). Original work in algebra—including cubic equations—was also done by the famous Persian poet Omar Khayyam (1048–1131).

Al-Khwārizmi also wrote *Concerning Hindu Numbers*, which explained the decimal place system and the figures we now call Arabic numerals, which originated in India. This decimal system was not widely used at first, but became increasingly popular in the following centuries.

Other Islamic mathematicians did extensive work in geometry, building primarily on Euclid's *Elements*. They developed the field of trigonometry as we know it, using ideas from Ptolemy's *Almagest* (which employed a system of chords) and the Indian *Siddhanta* (which introduced the sine function). Mathematicians created new trigonometric functions, documented them in numerical tables, and applied them to plane and spherical surfaces.

* * *

In medicine and anatomy, Islamic physicians built primarily on the work of Hippocrates, Dioscorides, and Galen, but they also made use of cures from India, Persia, and pre-Islamic Arabic sources.

The Persian polymath al-Rāzi (Latin name Rhazes, c. 854–c. 925) earned a reputation as the outstanding physician of his time. Considered a master clinician, he oversaw several hospitals (in Rayy and Baghdad), introduced new surgical procedures and remedies, and wrote dozens of medical books. He performed what has been called the first clinical trial, testing the effects of bloodletting on patients with meningitis.

Al-Rāzi's approach was overwhelmingly empirical, basing his conclusions only on observational data. Based on his comprehensive data collection, his studies of diseases became indispensable for other physicians. His book *On Smallpox and Measles* was eventually translated into Latin and other European languages, and went through forty editions between the fifteenth and nineteenth centuries. Al-Rāzi also wrote an encyclopedic twenty-volume work on medicine, titled *Kitab al-Hawi* ("Comprehensive Book"). Translated into Latin in 1279, this became the standard medical textbook in Europe for centuries.

The full extent of al-Rāzi's courage and intellectual independence is evident in his attitude toward religion. Al-Rāzi was born into Islam, but rejected it, along with all religion. He expressed particular hatred for all prophetically revealed religions: "How can anyone think philosophically while committed to those old wives' tales, founded on contradictions, obdurate ignorance, and dogmatism?"[82]

Next to al-Rāzi, the other dominant figure in the history of Islamic medicine is the philosopher and physician Ibn Sīna (Latin name Avicenna, 980–1037), who wrote the enormously influential *Canon of Medicine*. This was a detailed and systematic presentation of everything known about medicine and anatomy, and it became a classic on the subject. After its translation into Latin, it superseded al-Rāzi's encyclopedia and became the standard medical textbook in Europe.

In spite of explicit prohibitions on human dissection, a handful of physicians engaged in significant anatomical studies. In the ninth century, the translator Hunayn ibn Ishaq wrote a treatise on the eye, which included remarkably accurate anatomical drawings. Later physicians documented the human circulatory and nervous systems.

The greatest medical discovery of the Islamic world was probably that of the pulmonary circulation, which carries blood in a loop from the heart to the lungs and then back to the heart. Galen had held that the heart delivers blood to the body (including the lungs) through an ebb and flow, not a circulation.

The pulmonary circulation was discovered by Ibn al-Nafīs (1213–1288), chief of physicians at the Almansouri Hospital in Egypt. His *Commentary on Anatomy of the Canon of Ibn Sīna* contained this statement on the pulmonary circulation:

> When blood in [the right] cavity [of the heart] becomes thin, it must be transferred into the left cavity, where the pneuma is generated. But there is no passage between these two cavities, for the substance of the heart separating them is impermeable. It neither contains a visible passage, as some people have thought, nor does it contain an invisible passage which would permit the passage of blood, as Galan thought. . . . It must, therefore, be that when the blood has become thin, it is passed into the arterial

vein [our pulmonary artery] to the lung, in order to be dispersed inside the substance of the lung, and to mix with the air. The finest parts of the blood are then strained, passing into the venous artery [our pulmonary vein] reaching the left of the two cavities of the heart, after mixing with the air and becoming fit for the generation of pneuma.[83]

This discovery of Ibn al-Nafis did not become well-known in the Islamic world, and was rediscovered early in the twentieth century by a researcher looking through manuscripts in a library in Berlin. It is possible that the supposed European discoverer of the pulmonary circulation, the Italian Realdo Colombo (1510–59), may have known this discovery of al-Nafis.

* * *

In the field of astronomy, Muslims based their work on Ptolemy's *Almagest*. Muslims made progress in astronomy in three ways: (1) making observations to check and correct Ptolemy's parameters for the planetary models; (2) improving Ptolemy's models so that they made more physical sense; (3) setting up staffed observational observatories.

The observations designed to check Ptolemy's parameters began during the reign of the caliph al-Ma'mun, who initiated a program of observation at observatories in Baghdad and Damascus in order to check and improve the parameters of Ptolemy's system. One of the more famous astronomers for this type of work was the Syrian al-Battāni (Albategni, 858–929), who calculated new values for solar and lunar movements and the inclination of the ecliptic, calculated improved parameters for the motions of the planets, and created a corrected star catalogue. Copernicus and Kepler would later cite the work of al-Battāni (in Latin translations).

Some of the more significant astronomical discoveries were made centuries later, to the East, associated with the "Marāgha school." After the Mongol invasion of the Middle East destroyed the Abbāsid dynasty in the thirteenth century, the Persian astronomer Nasir al-Din al-Tūsi

(1201–74) accepted a position under the leader Hulahu Khan (grandson of Genghis Khan), who firmly believed in the value of astrology. Al-Tūsi persuaded Khan to fund the building and continued support of an observatory in Marāgha (in present-day northwestern Iran). In continuous operation for at least fifty years, this observatory would inspire the building of observatories in Samarkand, Delhi, and Jaipur.

In Ptolemy's system, a planet moves in a circular motion around an epicycle, with the center of the epicycle itself moving around a circle (the deferent). The center of the deferent is not at the earth but offset from it. Ptolemy also found the need for a further element: the equant. The equant is a point in space on the opposite side of the earth from the deferent. From the point of view of the deferent, the center of the epicycle moves along a circle, speeding up and slowing down. The equant is the point around which the center of the epicycle moves at uniform angular velocity.

Ptolemy usually discussed his system as a merely mathematical abstraction, without relating items like the epicycles, deferents, and equants to actual physical elements. But Greek thinkers such as Aristotle (and Ptolemy himself in a few places) described the celestial motions as caused by the uniform rotational movements of interlocking crystalline spheres in which the planets are embedded.

It was hard to see how an abstraction like the equant could relate to any such spheres, or have any apparent physical meaning. So numerous Islamic astronomers tried to find physical mechanisms that could give equivalent mathematical motions, but which, unlike the system of Ptolemy, would make more physical sense.

This led al-Tūsi to invent what became known as the "Tūsi couple"—a mechanism of converting a combination of two uniform circular motions into a back-and-forth straight-line motion. When, in the sixteenth century, Copernicus created his heliocentric system, he incorporated the Tūsi couple into his system.

Ibn Shātir (c. 1305–c. 1375) achieved what was considered the crowning achievement of Islamic astronomy. He produced models for the moon and planets that used double epicycles, and finally provided a physically plausible (and mathematically equivalent) substitute for the equants used by Ptolemy. Copernicus would later incorporate all of the mathematical elements of Ibn Shātir's system into his own.

* * *

Optics—the geometrical study of vision—was another field that inspired significant research in the Islamic world; and it was especially popular among astronomers. Their primary sources on this subject were Euclid, Aristotle, and Ptolemy.

The Babylonians, Egyptians, and Assyrians had used polished quartz to make simple lenses, and Greeks such as Euclid started laying down the basic principles of geometric optics. These included the idea that light travels in straight lines, the ways that light reflects from plane mirrors, and (from Ptolemy) a general understanding of refraction (the bending of light such as when it goes from air into water).

The Greeks also speculated about the nature of the process by which we see physical objects. Euclid and Ptolemy advocated the "extromission" theory of light: Our eyes emit light rays (in straight lines emanating like a cone) that illuminate the objects we see. Aristotle, by contrast, advocating an "intromission" theory, believed that light comes to our eyes from the objects we look at, with the image of the object entering the eye instantaneously. Others advocated a combination of the two, with light emitting from the eye, hitting the seen object, and then returning to the eye.

Al-Kindi (the first major philosopher of Islam) wrote a sharp critique of the extromission theory, pointing out the absurdity of light having to go all the way from our eyes to the stars, in order for us to see them. This critique was taken up and expanded upon by Ibn al-Haytham (Alhazen, c. 965–1039), whose famous *Kitab al-Manazir* ("Book of Optics") would become the standard text on optics for future generations. Both al-Kindi and al-Haytham claimed that light does not radiate from a luminous object as a single unit; instead, it radiates in all directions from each point on the surface of the object, independently of light radiating from other points.

Al-Haytham undertook a systematic experimental investigation of the properties of light and how it interacts with different types of objects. This included experiments with sunlight, candlelight, and reflected, refracted, and colored light, using sighting tubes, darkened chambers, and lenses. He described in detail his improved version of Ptolemy's instrument for measuring refractions, and he explained how he used it to study air-water, air-glass, and water-glass refractions for planes and curved surfaces. In general, his experiments served to confirm common-sense ideas such as that light travels in straight lines, and that light rays from different directions can travel independently through the same point in space.

Al-Haytham also made effective use of his mathematical expertise in his exact solutions to what became known to seventeenth-century Europeans as "Alhazen's problem": From any two points in front of a reflecting surface (which may be curved, as in spherical, cylindrical, or conic), find the point (or points) on the surface at which the light from one point will be reflected to the other.

Building on his observations, al-Haytham developed a comprehensive theory about how we see external objects, explaining how light is emitted from each point on the visible body. When the light enters the eye, it is refracted in such a way so that the eye's vitreous humor could sense the resulting image (the "form" of the object seen). He stopped short of considering the eye as a camera obscura or lensed camera. (Johannes Kepler would later make this discovery in the sixteenth century.)

Because of his extensive use of experiments and mathematical analysis, al-Haytham has been referred to as the "first true scientist." Al-Haytham's book of optics would later be translated into Latin and avidly studied by Europeans such as Kepler and Descartes, who considered it the chief authority on the subject of optics.

Two other experimental discoveries in optics are worth noting. Ibn Sahl (fl. 985), in Baghdad, did an experimental study of the refraction of light passing from one transparent medium to another, through straight and curved surfaces. By measuring the angles of incidence and refraction, he arrived at the geometrical equivalent to the modern law of refraction (known today as Snell's law).

The other optical discovery involved the nature of the rainbow. Al-Fārisi (1267–1319), of the Marāgha circle, was familiar with al-Haytham's book on optics (and his experiments). Inspired by Ibn Sīna's theory that it is the sunlight's reflection from individual water droplets that creates the rainbow, he decided to do his own experiment. He created a glass sphere and filled it with water, then shown rays of solar light on it. By this means, he was able to determine that both reflection and refraction were necessary to produce the rainbow. When light enters the "drop" it is refracted, then it is reflected at the back of the drop, then refracted again on leaving the drop. This creates the primary rainbow. When the light was reflected twice, this produced the secondary rainbow. (This was the very same experiment that was done independently by Theodoric of Freiberg in Europe around the same time; significantly, Theodoric had also been inspired by al-Haytham's book on optics.)

* * *

The preceding pages have given an overview of the scientific achievements of the Islamic Golden Age. We can see that the scientists of this era got their intellectual starting points by studying the Greeks. Based on these starting points, they made real advances. Historian Toby Huff argues that, "Considered altogether, in mathematics, astronomy, optics, physics, and medicine, Arabic science was the most advanced in the world."[84]

But this progress did not last. After three centuries of intellectual productivity, the Islamic world went into a decline from which it has never recovered. In order to understand why this occurred, we must look at the relation between reason and faith in the Islamic world.

Islamic Faith versus Reason and Philosophy

IN SPITE OF the genuine scientific achievements listed in the previous section, it would be a mistake to assume that science was deeply integrated into Islamic culture. Muslims emphasized a distinction between the "foreign" or "intruding" or "alien" sciences, such as philosophy, physics, astronomy, and medicine, and the "Islamic" sciences, which concerned themselves with the Quran, the Hadith (the sayings attributed to Muhammad), and Islamic law. Two aspects of the "foreign" sciences made them alien to Islamic culture: (1) They "intruded" from Greece and were not indigenous to Islamic culture; (2) More fundamentally: they employed the method of reason, not faith.

While there was a long-lasting tradition of studying the "foreign" sciences, this tradition was always centered on a relatively small number of wealthy and powerful individuals for patronage. Science and philosophy were rarely taught in the established schools—the madrasas—which focused on the "Islamic" sciences.

The overall Muslim attitude toward reason and science is evident in its approaches to both philosophy and theology. Consider three of the prominent "philosophers" of the Islamic world: al-Kindi, al-Farābi, and Ibn Sīna. Al-Kindi held that God created the universe out of nothing and that everything in the Quran is true and reconcilable with reason. Al-Farābi wrote that reason provides us with the truth because God, as creator and controller of the universe, is the source and archetype of reason. Ibn Sīna, esteemed as a great scholar of Aristotle as well as of medicine, held that universals exist in the mind of God, and that God uses these as models for creating things in the world. Given the central role that Islam played in their thought, it is arguably more accurate to classify these thinkers as theologians than philosophers, "theology" being the use of deductive logic to infer aspects of reality from faith-based starting points.

Now let us turn to the theologians.

The Abbāsids—the supporters of the translation movement and the House of Wisdom—also supported the first fully developed theological school in Islam, known as the Mu'tazilites. The Mu'tazilites used Greek philosophical concepts and logic to explicate Islamic teachings. They championed what they called "reason," but for them this meant the use of deductive logic starting from faith-based premises. Their "reason" was not an inductive reason that looks at the world and bases its conclusions on observable evidence.

Saying that "the mind can know things, as distinct from having opinions about them" and that "objective reality exists," the Mu'tazilites provided elements of a philosophical foundation that could, in some sense, help to justify natural science:

> The concept of an inherent nature in things . . . means that God, though he is the First Cause, acts indirectly through secondary causes, such as the physical law of gravity. In other words, God does not immediately and directly do everything. He does not make the rock fall; gravity does. God allows some autonomy in His creation, which has its own set of rules, according to how it was made. As Mu'tazilite writer and theologian 'Uthman al-Jahiz (776–869) stated, every material element has its own nature. God created each thing with a nature according to which it consistently behaves. The unsupported rock will *always* fall where there is gravitational pull. These laws of nature, then, are not an imposition of order from without by a commander-in-chief, but an expression of it from within the very essence of things, which have their own integrity.[85]

This distinction between primary and secondary causes is one that we will see in the Christian world as well. This distinction offers the scientist a semi-plausible "middle-way" between the *primacy of existence* (which his scientific work requires) and the *primacy of consciousness* (which his religious belief requires). This allows the scientist to split his soul into two halves to avoid thinking about religion while he is doing science. It withdraws God from an active role in our lives.

Why is this distinction only semi-plausible? Because if God is truly all-powerful, then He can decide to violate the laws at any moment, as frequently as His whims dictate, and He can make any senselessly bizarre thing occur.

As supporters of a deductive reason who were influenced by Plato (and Aristotle, to a lesser extent), the Mu'tazilites came into conflict with Islamic traditionalists (such as Ibn Hanbal) on numerous issues, such as free will, the "created" nature of the Quran, and the interpretation of the literal Quran. The Quran's references to God's "hands" and "face" and God's "sitting on the throne" generated much debate. The Mu'tazilites interpreted these references as metaphors. But traditionalists held that, as part of the revealed Quran, these terms must be literally correct. One prominent traditionalist said, "The sitting is known, its modality is unknown. Belief in it is an obligation and raising questions regarding it is a heresy."[86]

The relatively pro-reason view of the Mu'tazilites came under sustained attack from a rival school of theology that emphasized the omnipotence of God's will. The founder of this school of theology was Abu Hasan al-Ash'ari (873–935). Seeking to return to a more traditionalist approach to Islam, the Ash'arite view focused on God's uncompromising omnipotence and will: "Allah does what He wills." (Quran 14:27).

> The Ash'arite argument reduced God to His omnipotence by concentrating exclusively on His unlimited power, as against His reason. God's "reasons" are unknowable by man. God rules as He pleases. Allah had only to say "be" in order to bring the world into existence, but He may also say "not be" to bring about its end—without a reason for doing either.[87]

The Ash'arite turn back to traditionalist Islam even led to the explicit dismissal of the value of physics. The Ash'arite Ibn Khaldun put it this way:

> We must refrain from studying these things . . . since such restraint falls under the duty of the Muslim to not do what does not concern him. The problems of physics are of no importance for us in our religious affairs or our livelihoods. Therefore we must leave them alone.[88]

In reviewing the main branches of Islamic philosophy and theology, it is clear that the Greek perspective of reason based on looking at the world was never popular. Almost all Islamic "philosophy" uses God and his omnipotence as a starting point. The Islamic theologians share this tenet, and philosopher Andrew Bernstein explains:

> The Mu'tazila held that God is reason, His creation is rationally intelligible, and consequently, reason—by which they meant deductive logic from the definition of God—is the final arbiter of theologic inquiry. Al-Ash'ari and his supporters start with the Koran and the Hadith; they hold that reason can, in some cases, explicate scriptural teachings; and they hold that the holy texts are, in all cases, to be accepted, never doubted. Ibn Hanbal and his followers embrace the holy texts literally and eschew all theology.[89]

Bernstein concludes that "Philosophy, as conceived by the Greeks— observation-based rationality applied to life's most fundamental questions—never openly existed under Islam."

Al-Ghazāli as the Culmination of Islam's Attack on Reason and Science

THE MOST IMPORTANT theologian of Islam was al-Ghazāli (Algazel, 1058–1111), who was deeply influenced by the Ash'arites. In his *Incoherence of the Philosophers*, al-Ghazāli not only attacked the value of philosophy and science, he directly attacked the idea of cause and effect, which is fundamental to science:

> The connection between what is habitually believed to be a cause and what is habitually believed to be an effect is not necessary. . . . For example, there is no causal connection between the quenching of thirst and drinking, satiety and eating, burning and contact with fire. Light and the appearance of the sun, death and decapitation, healing and the drinking of medicine, the purging of the bowels and the using of a purgative, and so on to [include] all [that is] observable among connected things in medicine, astronomy, arts, and crafts. Their connection is due to the prior decree of God, who creates them side by side. . . . [It] is within [divine]

power to create satiety without eating, to create death without decapitation, to continue life after decapitation, and so on to all connected things. . . .

Our opponent claims that the agent of the burning is the fire exclusively; this is a natural, not a voluntary agent. . . . This we deny, saying: The agent of the burning is God, through His creating the black in the cotton and the disconnection of its parts, and it is God who made the cotton burn and made it ashes either through the intermediation of angels or without intermediation. For fire is a dead body which has no action, and what is the proof that it is the agent? . . . [O]bservation proves only simultaneity, not causation, and, in reality, there is no other cause . . . but God.[90]

So here again, we see the *primacy of consciousness*, laid bare.

Al-Ghazāli's most famous critic was the prominent physician and philosopher Ibn Rushd (Averroës, 1126–98). In his book *The Incoherence of the Incoherence*, Ibn Rushd systematically answered al-Ghazāli and defended the value of philosophy and deductive reason. Through his extensive commentaries on the writings of Aristotle, Ibn Rushd established himself as the greatest Aristotelian scholar of the Islamic world.

Ibn Rushd worked systematically to adapt Aristotelian ideas into an Islamic framework. He agreed with Aristotle that the universe is in some sense eternal (not created ex nihilo or at one point in time), although he held that in another sense, God had created the world. Like the Mu'tazilites, he ultimately saw God as the source of everything, and he accepted the Quran as infallible (although often requiring careful interpretation). The last of the major Islamic philosophers, Ibn Rushd was highly influential on later Christians and Jews, but within the Islamic world he was ignored. The clear victor in this battle was al-Ghazāli, who is still revered by Muslims throughout the world.

Science Fades Away and European
Discoveries are Ignored

THE VICTORY OF al-Ghazāli and the fundamentalists did not bring an immediate end to Islamic science, which slowly declined in the following several centuries. But the overall effect was apparent: The forces of religious conservatism now had the upper hand, and science was in retreat. The study of the natural world had lost its respectability.

By the sixteenth and seventeenth centuries, the full extent of the Islamic introversion and what Toby Huff has called a "curiosity deficit" becomes apparent when we look at the Islamic reaction to two key European inventions: the telescope and the printing press.

The telescope was arguably the most important invention of the Scientific Revolution. When Galileo turned his telescope to the sky in 1609, his discoveries triggered an avalanche of ideas and further discoveries, ultimately overthrowing the classical view of the universe. The Islamic reaction to the telescope was notably different. According to historian Toby Huff:

> Telescopes had been widely disseminated across the Ottoman Middle East, among the Safavids, and in the Mughal Empire within a decade or two of their invention in Europe in 1608–9 and the appearance of Galileo's revolutionary revelations in *The Starry Messenger* of 1610. Yet, the telescope's arrival in Muslim lands—in Mughal India, the Ottoman Empire, and elsewhere—hardly created a stir. No new observatories were built, no improved telescopes were manufactured, and no cosmological debate about what the telescope revealed in the heavens have been reported.[91]

When the telescope came into use in the Islamic world, Muslim astronomers ignored it:

> There is no indication that Middle Eastern scholars used the telescope for astronomical exploration, nor did they make any replicas of the Western-invented telescopes that arrived in the Muslim

world. . . . Neither do we have any indication that Muslim scholars were in search of or open to new theoretical ideas in astronomy. Indeed the shift to a Copernican and Newtonian worldview was greatly delayed [in the Middle East].[92]

Unlike the telescope, the printing press was not a scientific instrument, but it was crucially important to the widespread dissemination of knowledge in Europe. Gutenberg printed his Bible in the 1450s, and soon after this, presses appeared in all the major cities of Europe. The following century saw an explosion of printed books on a wide variety of subjects. But the story in the Islamic world was different:

> With its wealth, scholastic traditions and urban comforts, Islam, you might think, was a perfect seedbed for the printing press. The Muslims had paper; they had ink. . . . Knowledge of printing came with Jewish refugees fleeing from persecution in Spain to Constantinople. . . . In 1493 Jewish refugees (from Spain) in Constantinople produced the first books in Hebrew (bibles and secular books). A Qur'an was printed in Arabic in Italy by 1500. A generation later, in 1530, . . . the grandson of Israel Nathan Soncino, founder of the great Jewish/German/Italian publishing family, set up in Istanbul, and later moved to Cairo. There is no way that an educated Muslim could not have known about printing or its potential. And yet from Muslim traders, scholars, administrators, not a flicker of interest, or downright hostility. When the first printing works was established in Istanbul in 1729 . . . [the printer] managed to print just seventeen titles before religious opposition became so intense that the press was closed down in 1742.[93]

The use of the printing press did not become widespread in the Islamic world until the mid-nineteenth century, four hundred years after it had become established in Europe. Why did it take so long?

The main reason for Islamic opposition to printing seems to be the "objective" nature of the printed word. Historian Francis Robinson has argued that "Printing attacked the very heart of Islamic systems for the transmission of knowledge; it attacked what was understood to make knowledge trustworthy, what gave it value and authority."[94]

What supposedly made knowledge trustworthy was its direct oral transmission from one person to another, as the Quran was transmitted:

> The Quran was always transmitted orally. This was how the Prophet transmitted the message he had received from God to his followers. When, a few years after the Prophet's death, these messages came to be written down, it was only as an aid to memory and oral transmission. This has been the function of the written Quran ever since.[95]

> Printing, by multiplying texts willy nilly, struck right at the heart of person to person transmission of knowledge; it struck right at the heart of Islamic authority.[96]

In the form of a printed text, a work has a type of objectivity; as a physical object, it has a certain independence from anyone's consciousness. It can be studied without explicit regard for who created it, by focusing on verifying it against reality.

But in the Islamic view, knowledge was supposed to be passed directly from one person to another, in the form of vocal recitations. If there was no direct tie between the person of the source and the reader, then the knowledge was suspect, in doubt. This is an authoritarian view of knowledge itself. This view is also a sign of a deeper philosophic orientation: *the primacy of consciousness*—in this case, the consciousness of the originator of the text.

* * *

The *primacy of consciousness* orientation explains much more than just the Islamic aversion to the printing press; it explains the ultimate failure of science in the Islamic world.

The achievements of Islamic science came from the importation of the *primacy of existence* from Ancient Greece. But unfortunately, the philosophy of Aristotle was never to penetrate deeply into the Islamic world. Even the more pro-logic Mu'tazilites were not real supporters of observation-based reason.

The theology of the Ash'arites was an unusually consistent form of the *primacy of consciousness*. By stressing God's omnipotence, they stressed the primacy of God's consciousness, and completely undercut the foundations of reason and science.

The ultimate result is clear. In the words of the twentieth-century Pakistani physicist Pervez Hoodbhoy, "science in the Islamic world essentially collapsed. No major invention or discovery has emerged from the Muslim world for over seven centuries now."[97] The victory of the *primacy of consciousness* was indeed complete.

CHAPTER SIX

Europe's Discovery of Aristotle Leads to the Renaissance

As WE SAW in Chapter Three, in the early Middle Ages, the dominant philosophy was that of Augustine (and thus Plato). According to this philosophy, there are two worlds: the heavenly realm above and this low, physical world below. We are pulled down by this world and the flesh. We are "deformed and squalid" and "tainted with ulcers and sores."[98] Our only significant source of ideas and values is faith and the supernatural. Accordingly, during this period, people directed all their intellectual activity towards the other world. They performed no science of note.

From the sixth century through the beginning of the twelfth century, the only books of Aristotle known to the Latin West were his books of logic (the *Organon*) minus the *Posterior Analytics* (his work on scientific demonstration). But this was about to change.

The Reconquista Leads to a Translation Movement from Arabic into Latin

As RECOUNTED IN the last chapter, Muslim armies conquered the bulk of the Iberian Peninsula early in the eighth century. The Umayyad dynasty ruled Muslim Spain (Al-Andalus) from 756 to the early 1000s. (The Umayyads were succeeded by the Muslim Berber dynasties the Almoravids and the Almohads.) The Umayyads built and supported an extensive library in Córdoba, and patronized intellectuals such as court physicians and astronomers, competing with the Abbasids center in Baghdad to attract the best thinkers.

Starting in the tenth century, Christian forces exerted a prolonged effort, involving numerous wars, to regain control over the peninsula. Historians now call this multi-century effort the *Reconquista*, meaning "re-conquest." Christian forces had taken Toledo and Lisbon by 1100, and Córdoba and Seville by the first half of the thirteenth century. In 1492, the last of the Muslim kingdoms in Iberia was defeated, bringing the entire Iberian Peninsula under the Christian rule of Ferdinand and Isabella.

The conquering Christians discovered a culture that—to their credit—they recognized as more advanced than their own; this new culture was familiar with many valuable books unknown to the West, so they began to translate these books into Latin. Gerbert of Aurillac (c. 945–1003), who would be elected as Pope Sylvester II in 999, was one of the first Europeans to recognize the value of the translated Arabic texts. When he was archbishop of Reims, he lectured on the logical writings of Aristotle and others. His main intellectual interest was the classical mathematical quadrivium (arithmetic, geometry, astronomy, and musical theory).

A number of Arabic astronomical and mathematical texts made their way to Catalonia in Christian-controlled northern Spain, and into the hands of Gerbert when he was visiting the area. When Gerbert returned, he wrote about the ideas he had learned, including the Arabic numerals, and his writings generated extra interest in the translations. This translation movement lasted through the twelfth century and into the thirteenth century, and as a result, beginning in the 1150s, Latin editions of Aristotle, translated from Arabic, started becoming widely available in libraries.

Suspicious that the works of Aristotle and others had been corrupted in the process of multiple translations, Europeans then began looking for texts in the original Greek, which could often still be found among scholars from Constantinople who had preserved many Greek works. Eventually, translations were made directly from Greek into Latin. (The flow of Greek manuscripts from Constantinople ended after the city fell to the Ottomans in 1453.)

Aristotle's ideas, in one form or another, became almost ubiquitous in European intellectual life. One sign of this is the fact that competing Christian factions even cited Aristotle in their arguments against each other. A case in point was the story of the Cathars. This anti-Catholic dualistic sect, based in southern France, achieved a high level of popularity in twelfth-century Europe. Cathar intellectuals explicitly relied on Aristotle's principle of non-contradiction to point out the problem of evil. A good and omnipotent God cannot be reconciled with the obvious existence of evil in the world. The Cathars argued that there must have been two creations, not just one. The good God created the world of the spirit, while at the same time, an evil God, equal in power, created the material world. Opposing the Cathar view, churchmen such as the Dominicans argued that Cathars were actually in fundamental conflict with Aristotle's view that there is only one universe, and that it is not evil.

The Cathar threat to Catholicism was not ultimately removed by arguments, but by force. Pope Innocent III's call for a crusade against the heretics in 1208 led to their brutal and bloody extermination. After the crusaders stormed the city of Beziers, the papal representative proudly declared that "nearly twenty thousand of the citizens were put to death, regardless of age and sex."[99]

Aristotle's This-Worldly Philosophy is Fully Rediscovered

THE FULL DISCOVERY of Aristotle's writings was a shock to the medieval mind: "The Aristotelian stance, with its unabashed admiration for the material world, its distaste for mystical explanations of natural phenomena, and its optimism about human nature, ran counter to centuries of otherworldly, ascetic Christian values and practices."[100] To the medieval Europeans, who had been surrounded by ignorance and otherworldliness, the extent of Aristotle's interest in and knowledge of the physical world was impressive and overwhelming.

But many in the Church disliked Aristotle for his almost anti-Christian philosophy. In addition to his overall this-worldly orientation, Aristotle had explicitly argued for several ideas that clearly contradicted Christian doctrine:

- The world has always existed; it is eternal.[101]
- There is no personal immortality; when a person dies, his personal soul ceases to exist.[102]
- Pride is not a sin; to the contrary, it is the "crown of the virtues."[103]

The interest in Aristotle developed primarily in universities. The universities had originated in cities such as Bologna, Paris, and Oxford during the twelfth century. They consisted of a single faculty of arts (which provided an undergraduate curriculum) and graduate programs in theology, medicine, and law. The Church had official control over the universities, but it rarely exercised tight control. Most university students had no plans to enter the Church. The universities provided a basic education in classic Greek & Roman texts—an education that became universal throughout Europe.

The intellectual center of Europe was the University of Paris, and during the thirteenth century, it was home to a major conflict about Aristotle's role in the curriculum. In 1210, the Parisian Synod (a council of bishops) issued a decree that "no lectures are to be held in Paris either publicly or privately using Aristotle's books on natural philosophy or the commentaries, and we forbid all this under pain of excommunication."[104] In 1215, this ban was renewed. In 1231, Pope Gregory IX issued another renewal of the ban, specifying that Aristotle's books on natural philosophy were not to be used until they had been "examined and purged of all suspected error."[105] But these bans were not effective, and by the 1240s, a number of teachers were lecturing on Aristotle's works. In spite of Pope Gregory's ban, no one ever apparently created a purged version of Aristotle. During the time of these bans, Aristotle's works were openly studied at other universities, such as Oxford.

In spite of the bans, interest in Aristotle continued unabated, and in 1255 Aristotle's writings on natural philosophy became required reading for all faculty of arts students at the University of Paris: "By the second half of the thirteenth century, [Aristotle's] works on metaphysics, cosmology, physics, meteorology, psychology, and natural history became compulsory objects of study. No student emerged from a university education without a thorough grounding in Aristotelian natural philosophy."[106]

Why did Aristotle's philosophy become so popular? Ever since the Church Fathers (such as Justin Martyr, Clement of Alexandria, and Ambrose of Milan), Christian culture had a legacy of at least partial respect for pagan Greek philosophy. And in spite of the challenges it posed, Aristotle's philosophy offered a distinct value, as Richard Rubenstein relates in his book *Aristotle's Children*:

> The Aristotelian corpus, troubling though it might be, represented the most comprehensive, accurate, well-integrated, and satisfying account of the natural world that medieval readers had ever encountered. Here one could discover a method of rational inquiry . . . and the ways of spotting illogical ("sophistical") arguments. Here nature's mysterious secrets were disclosed: how fish sleep; why a thrown object continues in motion after it leaves the hand; how the Milky Way was formed. From such particulars, the reader could ascend with the Philosopher to breathtaking generalities such as the types of possible knowledge and classifications of the sciences, the dynamics of natural change, and the nature of Being itself. Part of the work's appeal lay in its encyclopedic scope. Long known in Europe as a master of formal logic, Aristotle was now revealed as the master of all knowledge . . . whose ideas would henceforth become the starting point for almost all discussions of natural or social phenomena.[107]

Two key figures who helped ensure Aristotle's continuing influence were the Dominican friars Albertus Magnus (Albert the Great) and his student Thomas Aquinas.

Thomas Aquinas Creates a "Protected Area" for Reason

BORN TO A wealthy Bavarian family, Albert of Lauingen (c. 1200–80) studied the liberal arts at Padua, in Italy, joined the Dominican Order, and was ordained a priest. He went to the University of Paris c. 1241, and this is where he first discovered the recently translated writings of Aristotle. Albert wrote extensive paraphrases of almost all of Aristotle's works, and he soon got a reputation as a master of all knowledge.

From an early age, Albert was interested in the natural world, and his discovery of Aristotle further validated this interest and encouraged him to give it full expression. Given the time in which he lived, the scope of Albert's investigations is truly impressive. He studied falling objects, rainbows, stars, and the moon. He did experiments on the effects of sunlight in generating heat. He did alchemical experiments and isolated the element arsenic. He studied fossils as well as minerals and their properties. He did extensive comparative studies of the structures and parts of plants, making numerous original discoveries. He performed studies of animals, especially their reproduction and embryology. He did studies of insect mating. All of this interest in the natural world led to the reasonable accusation that he was neglecting theology.[108]

While teaching at the University of Paris, Albert encountered a brilliant student who shared his love of Aristotle, and who would go on to eclipse Albert's philosophic achievements.

Thomas Aquinas (1224/5–1274) was a brilliant Dominican friar. Originally from the Roman Campagna, he studied at the University of Naples, the first university founded independently of the Church. (Holy Roman Emperor Frederick II founded it to train government officials.) Here Aquinas learned about and studied Aristotle's natural philosophy (using Arabic translations and Islamic commentaries), which at this time was banned at most other universities. He later went to the University of Paris, where he studied under Albert, and he then traveled with Albert to Cologne to help with establishing a school.

Aquinas's masterwork *Summa Theologica* presents his extensive synthesis of Aristotelian philosophy with Christian theology. The most important aspect of his thought was his view of the status of reason:

> Thomas startled his contemporaries and initially earned the reputation of being a dangerous radical by extending the realm of reason deep into the territory of theology. . . . According to him, there are only three doctrines that cannot be proved by using natural reason: the creation of the universe from nothing, God's nature as a Trinity, and Jesus Christ's role in man's salvation. These truths, which cannot be deduced from experience, depend on faith alone. But other theological doctrines, even the existence of God, as well as the Creator's immateriality, perfection, goodness, knowledge, and other attributes, can be arrived at by using one's reason to analyze and generalize from observed data.[109]

Whereas Augustine had seen faith as the base of reason, and Tertullian had seen faith as contradictory to reason, Aquinas saw faith as merely a supplement to reason. When it came to the nature of reason and the physical world, Aquinas accepted Aristotle's views: Reason is an autonomous faculty, which we must use and obey. The physical world is knowable and real. Life is about living, not cursing our existence.

Both Albert and Aquinas made a formal and explicit distinction between revealed theology and natural philosophy—between the truths of faith and the truths of natural reason. Revealed theology started with faith in authorities, and used logic to develop further knowledge. Natural philosophy started with observation and logic, and used only reason to come to conclusions. Albert and Aquinas held that both types of knowledge were valid.

Aquinas held that there was an overlap between these two domains of knowledge. This overlap consisted of truths of theology that could be established entirely by natural reason, without any use of faith. Such truths included the existence of God and His basic attributes such as omnipotence and omniscience. Aquinas presented five famous proofs for the existence of God, arguing from observational evidence. The use of reason to establish such truths of theology, Aquinas called *Natural Theology*. Truths of faith, on the other hand, included such items as the Trinity, the Virgin Birth, the Trans-substantiation of the Eucharist, and the Assumption of Mary.

This distinction is evident in an oath imposed on the masters of the faculty of arts:

> The faculty of arts at the University of Paris, beginning in 1272, required all masters to swear an oath that they would not introduce theological matters into their disputations. . . . In sum, the disciplines of theology and natural philosophy have nothing to do with one another and should be kept apart.[110]

The worldly philosophy of Aristotle is evident in many aspects of Thomistic thought, though always modified to fit into a Christian context. Aquinas held that the goal of life is happiness and that in order to achieve complete happiness, one needs to achieve divine grace. But he also held that we are natural creatures and that in this life on earth, we can achieve a certain degree of happiness. In this world, reason is efficacious, and we should use our reason. For Aquinas, of the two paths to truth—faith and reason—reason is more important for guiding our actions in this world, although he held that they can never conflict.

Another issue that reveals the influence of Aristotle on Aquinas is the nature of the soul. Augustine, in true Platonic fashion, had viewed the soul as being trapped in the body as in a prison. Aristotle wrote that the soul is the *form* of the body, and that it ceases to exist after the death of the body. This was not consistent with Christianity, and Thomas held that after death, the soul exists in an *unnatural* state, and it only returns to a natural state at the final resurrection.

The innovations of Aquinas in expanding the scope of reason were unprecedented for a Christian thinker. But even Aquinas's expanded scope of reason was outdone by the "radical Aristotelians," led by an outspoken teacher at the University of Paris named Siger de Brabant. Siger presented Aristotle's ideas about nature and human nature without even attempting to reconcile them with Christian doctrine, in effect leading to two independent realms of truth: truths of reason and truths of theology.

Siger argued that, *from the point of view of reason*, the universe is eternal, since all effects have causes which are the effects of prior causes. But *from the point of view of theology*, the universe was created at a certain point in time. Siger also claimed that reason leads to the eternity of mankind and the rejection of personal immortality.

But when pressed, he acknowledged that when faith leads to different conclusions than reason, then faith is the only source of truth. In spite of this concession, Siger and his associates were plagued by accusations of heresy, and Siger himself was eventually murdered. In 1277 an official Condemnation listed 219 different propositions, all of which were expressly prohibited not only in the classroom but also in private discussion. This list included numerous ideas advocated by Siger. The "radical Aristotelians" were effectively silenced.

The Thomists, on the other hand, were not silenced at all. In the half-century after the death of Aquinas in 1274, his influence continued to increase. In 1278, the Dominican Order of friars officially adopted the theology of Thomism. In 1323, the Church canonized Aquinas as Saint Thomas. The Thomist synthesis of Christian theology and Aristotelianism became a dominant force in the Church for centuries. Aquinas brought Aristotle's ideas into the mainstream of Christian thought and boxed in theology, isolating it as a separate academic discipline. Given the previously dominant "handmaiden" view of science, this amounted to the liberation of reason and interest in this world.

Aristotle as a Continuing Force in the Universities

FROM THE END of the thirteenth century on, the universities hosted a steady sequence of Aristotelian teachers, and this continued until the seventeenth century. The study of Aristotle and other classical Greeks led to the emergence of a new type of scholar, who was primarily focused on the nature of human beings as natural creatures, living in this world. These scholars were the humanists. Humanism first developed in Italy, with figures such as Petrarch (1307–74), Boccaccio (1313–75), and Giovanni Pico della Mirandola (1463–94). Giovanni Pico della Mirandola's *Oration on the Dignity of Man* (1486) has been called the "manifesto of the Renaissance." Key humanists also included Marsilio Ficino, Leonardo Bruni, Niccolo Machiavelli, Desiderius Erasmus, and Michel de Montaigne.

Although they had an implicit Aristotelian influence, the humanists had eclectic philosophical interests. Many were attracted to Platonism, but others were drawn to Aristotelianism, Stoicism, Epicureanism, or Skepticism. The full recovery of Plato's writings in the fifteenth century led to an increased interest in Plato. But overall, Aristotle was much more widely read than Plato during the Renaissance. From the 1450s to around 1600, as many as four thousand editions of the works of Aristotle may have been published, compared to fewer than 500 comparable editions of the works of Plato.[111] A wide variety of publications presented Aristotle's ideas, including textbooks, summarizing tables, introductions, and commentaries.

Leaders of the Protestant Reformation such as Martin Luther urged a return to the roots of Christianity in the Platonic theology of Augustine and the other Church Fathers, so these leaders tended to reject Aristotle. Luther referred to Aristotle as a "damnable, arrogant, pagan rascal" who "has seduced and fooled so many of the best Christians."

> [T]his defunct pagan has attained supremacy; impeded, and almost suppressed, the Scriptures of the living God. When I think of this lamentable state of affairs, I cannot avoid believing that the Evil One introduced the study of Aristotle. . . . [H]is book on *Ethics* is worse than any other book, being the direct opposite of God's grace, and the Christian virtues; yet it is accounted among the best of his works. Oh! away with such books from any Christian hands.[112]

Many who shared this hatred of Aristotle attempted to remove Aristotle's works from Protestant universities, but these attempts were ineffectual. There was too much demand for Aristotle, even among Protestants.

The academic interest in Aristotle often did not follow the spirit of Aristotle's philosophy. There were Aristotelians like Cesari Cremonini (1552–1631) who considered Aristotle's writings to be the ultimate standard of truth in natural philosophy, ranking higher than any actual firsthand study of the natural world. Cremonini was infamous for his refusal to look through Galileo's telescope.

But, crucially, some Aristotelian teachers did have respect for the actual spirit of Aristotle's philosophy. One of the most prominent Aristotelians of the sixteenth century was Jacabo Zabarella (1533–89), who taught at the University of Padua shortly before Galileo studied there. Zabarella's treatises on logic, methodology, and natural philosophy were popular throughout Europe.

> Zabarella strongly emphasized observation of the external world as a source of knowledge, while stressing that reason—not Aristotle—is the ultimate foundation of valid knowledge. Such a position is clearly spelled out in the lecture introducing his course for the academic year 1585: "I will never be satisfied with Aristotle's authority alone to establish something, but I will always rely upon reason; such a thing is truly both natural and philosophical for us, and I will also seem to imitate Aristotle in using reason, for in fact he seems never to have put forward a position without utilizing reason."[113]

The Renaissance as a Rebirth of Reason and Interest in This World

AQUINAS'S LIBERATION OF reason led to the Renaissance. After centuries of obsession with the next world, the Renaissance was a rebirth of interest in *this* world, the world of nature and living, breathing human beings. This can be seen vividly in the humanism and the art of the period: The Renaissance artists learned how to carefully examine the physical world around them. Their artwork showed a new level of realism and detail, and the figures in their paintings came to life in all three dimensions. Historian Lisa Jardine gave her history of the Renaissance the very appropriate title *Worldly Goods*.

The Renaissance artists painted realistic landscapes and portraits, and competed with each other to paint the most vivid textures, colors, and shadows. They resumed the classical interest in nude figures, in both painting and sculpture.

The ancient Greeks had developed techniques for creating the illusion of three dimensions on a two-dimensional surface. These included foreshortening, aerial perspective, and a type of linear perspective. Aerial perspective is the technique (typically used in landscapes) whereby objects in the distance are shown as paler, less detailed, and bluer than nearby objects. Foreshortening is the technique whereby objects not parallel to the picture frame are reduced. These techniques can be seen in surviving Roman murals at Pompeii. Artists abandoned the use of perspective during the Middle Ages because of its otherworldliness.

During the Renaissance, these techniques were rediscovered. In the late thirteenth century, artists such as Cimabue, Duccio, and Giotto began to use foreshortening and a non-systematic form of linear perspective. Early in the fifteenth century, Brunelleschi developed a systematic technique of linear perspective, which Alberti described in his treatise *Della pittura* (1435–6). After this, almost all Italian painters studied perspective. These techniques gave the Renaissance artists new power and freedom, allowing them to depict a far greater range of subjects realistically.

Renaissance artists also worked on developing new techniques for different painting materials. They often experimented with new materials and new ways of using materials, such as new ways of painting effectively on walls. They embraced a spirit of progress and self-improvement:

> It is likely that men such as Ghiberti and Brunelleschi saw themselves not just as artists but also as scientists . . . adding by progressive experiments to the sum total of human knowledge. Many of the great paintings of the time were demonstrations of what could be done and how to do it. Patrons knew this, and

encouraged it. Each time they commissioned a master, they were striving to help him push forward the frontier of knowledge and skill a little further—or in some cases a lot further. It was the true spirit of the Renaissance.[114]

This spirit was in full expression in the work of Leonardo da Vinci, in his observational drawings of anatomy and nature, his controlled experiments in water flow, his medical dissections, and his systematic studies of movement and aerodynamics. Although Leonardo's scientific writings were lost and did not influence the science of the following centuries, his scientific interests are a clear sign of the continuity between the new observational outlook and the development of science.

The new realism of the Renaissance extended even to literature, which became more concerned with the real lives of real human beings. Geoffrey Chaucer's *Canterbury Tales* (1386–1400) was written using a unique literary framework: a Canterbury pilgrimage to the shrine of St. Thomas Becket. Historian Paul Johnson notes

the vivid directness with which Chaucer brings out character, both in describing his pilgrims and within the tales they tell. It is the literary equivalent to the formulation of the laws of perspective and foreshortening by the artists of Florence. These men and women jump out from the pages, and live on in the memory.[115]

Another outgrowth of Renaissance naturalism was the "natural magic" tradition, in which people looked for "occult" (or "hidden") connections in nature—links not perceptually obvious or visible. Examples of such connections included the attraction of iron to magnets, the ability of sunflowers to follow the sun across the sky through the day, the Moon's effect on the tides, and the effects of medicines. These connections were often thought of as "sympathies" or "antipathies," in effect anthropomorphizing natural phenomena. In order to find these connections, people often looked for distinctive "signs" that God had placed on things. For example, the fact that a walnut looks like a brain was considered evidence that it might be a medicine for the brain. Many believed that grasping these hidden connections could be of great practical value, but they did not yet have a valid method for discovering or understanding them.

Another sign of the worldliness of the Renaissance is the emergence of educated men who bridged the gap between scholarly writing and practical trades. The self-educated French potter Bernard Palissy contributed to chemistry, geology, forestry, and agriculture. The humanist professor Georg Agricola, authored *De Re Metallica* (1546), a systematic treatise on mining and metallurgy that became the standard work on the subject for the following century. Both Agricola and Palissy would become inspirational figures for Francis Bacon. While not a scientist himself, Bacon would write much on scientific method that would be influential during the Scientific Revolution. He was also impressed by the usefulness of inventions such as the compass and the printing press, and by the Renaissance natural magic tradition insofar as it was focused on developing practical knowledge of nature.

The central invention of the Renaissance was Gutenberg's printing press, which actually required three different inventions: the moveable type, the ink, and the press. The first printed works that Gutenberg officially produced were published in the 1450s. Printers used the new press to print broadsides, pamphlets, and many different types of books. Soon printing presses were being built throughout Europe, and they were kept busy. By 1500, more than 25,000 different books had been published in Germany alone. Gutenberg's development of the printing press in the 1450s made possible the inexpensive reproduction of texts, greatly facilitating the rapid dissemination of knowledge.

Several other factors also contributed to the culture in which the Scientific Revolution would occur. The sixteenth-century Reformation had splintered the Church into numerous, warring camps, ending the monolithic hegemony with which the Church had dominated the European intellectual scene. Starting in the fifteenth century and extending through the sixteenth, a series of voyages of discovery, aimed at finding new trade routes, yielded an explosion of information about exotic places. The sea voyages of Prince Henry the Navigator, Bartholomew Diaz, Vasco da Gama, Columbus, and Magellan led to an awareness of new plants, animals, minerals, medicines, peoples, languages, and ideas. These factors all contributed to a cultural environment in which people were open to new ways of looking at the world. The stage was set for the Scientific Revolution.

* * *

IN THIS CHAPTER, we have seen Europe's discovery of Aristotle's this-worldly philosophy and its necessity for the Renaissance. In one respect, the Renaissance was a complete accident; Aquinas certainly did not see it coming. If he could have seen the full consequences of his embrace of Aristotle—if he could have seen the full worldliness of the Renaissance and the resulting scientific and industrial revolutions—he would probably have burned his own books.

The conflict over Aristotle during the thirteenth century can be seen as a conflict between the two metaphysical outlooks described in Chapter Four. As Aristotle entered Christian culture, people were turning away from the *primacy of consciousness* and toward the *primacy of existence*.

The 1277 Condemnation (of 219 articles) contains one (number 150) that embodies the central conflict. This article condemns the idea "That on any question, a man ought not to be satisfied with certitude based upon authority."[116] This article embodies the central conflict of the time. Should we accept ideas on the basis of faith in religious authorities, or should we do our own reasoning, based on our own observations, to come to our own conclusions? Article 150 advocates accepting the ideas of religious authorities without question. This placement of a mind (the authority) over reality (one's own observations and reasoning) is a clear form of the *primacy of consciousness*.

CHAPTER SEVEN

Galileo Versus the Church

T HE MOST INFAMOUS episode of conflict between science and religion is the case of Galileo's clash with the Church. This was a clash between the greatest scientist of the early Scientific Revolution and the greatest symbol of religion in Europe.

This chapter will present an overview of the events leading up to and including Galileo's trial and punishment. In the final section, we will see how modern historians have turned against the view that this clash was between science and religion, and we will see why they have rejected the most essential point of this story.

The Copernican Revolution

AT THE BEGINNING of the sixteenth century, the Ptolemaic system was still the foundation for all astronomy. The Muslim astronomers had made improvements to Ptolemy's system, but these were not dramatic changes. No one was ready to radically rethink Ptolemy's earth-centered system.

Early in the sixteenth century, Polish canon Nicolaus Copernicus developed a radically new system of the cosmos, published in his *De Revolutionibus* in 1543. Inspired in part by the sun-worshipping views of Aristarchus and Pythagoras, Copernicus proposed two new ideas: (1) The sun is fixed and unmoving at the center of the universe. (2) The earth participates in two motions: it rotates on its own axis, and it moves in a circle around the sun.

With his proposed system, Copernicus addressed one of the most disturbing mathematical issues that astronomers had with the Ptolemaic system. In order to account for the changes in speed of the planets as they move in their circles, Ptolemy had created an off-center point called the equant, which regulated the speed at which the planet moves. (The equant was the point at which the angular rate of change is constant, but beyond this, it has no physical meaning.) If the epicycles had seemed physically implausible, the equant was doubly so.

By rearranging the motions of the heavenly bodies, Copernicus was able to eliminate Ptolemy's equants completely. Moreover, whereas Ptolemy needed epicycles to explain the retrograde movements of the planets, Copernicus did not. (Copernicus did include epicycles in his system, and Kepler was later able to eliminate all epicycles by positing the even more radical notion that celestial bodies move in ellipses, not circles.)

Copernicus's system had other advantages as well: it explained why Mercury and Venus, unlike the other planets, are always viewed as being close to the sun. Ptolemy's system had no explanation for this. But it was explained by the Copernican system, since the orbits of Mercury and Venus are inside that of the earth. Ptolemy's system also exhibited another curious pattern:

> [T]he epicycle motions of Mars, Jupiter, and Saturn are correlated with the sun's orbit in such a way that the retrograde motions of these planets always occur when the sun is on the opposite side of the Earth. This is a striking coincidence in Ptolemy's theory: What does a planet's retrograde motion have to do with the sun?[117]

In the Copernican system, this was easily explained by the fact that these planets have orbits outside that of the earth.

Copernicus circulated his ideas to friends in a manuscript in 1514, but he was reluctant to publish, for fear of persecution. He knew that some would point to Biblical passages saying the earth was fixed. But the most common objection to Copernicus was: Why don't we perceive the earth's motion? Why do objects thrown up in the air come down in the same spot? If the earth is rotating with a period of twenty-four hours, not to mention circling the sun at high speed, then it seems that we should notice that we're moving. We should be continuously buffeted by strong winds; a rock thrown straight up in the air should come down far away. But we don't see anything like this.

A further problem with the theory involved a phenomenon known as *stellar parallax*, which is predicted by Copernicus's theory. Consider what happens when you stand at one goal-line of a football field and look towards the opposite end, focusing on the goalpost and bleachers. Now suppose that you walk along the goal-line and keep looking across the field. You will see the goalpost shift in position relative to the bleachers.

If the earth is moving in an enormous circle (its orbit) around the sun, the constellations should look different depending on where we are in that circle. The problem is that no one was able to measure any such changes. Copernicus explained this result by supposing that the stars are so distant as to make such changes too small to notice. Given the enormous stellar distances required for parallax to be unobservable, this was not a convincing argument for many of his contemporaries.

Copernicus's friends urged him to publish his theory, but he postponed this until he was on his deathbed. In 1543 he handed his book, *On the Revolutions of the Celestial Spheres*, off to the Lutheran theologian Andrew Osiander, to handle the publication. Without the knowledge of Copernicus, Osiander added an unsigned foreword, saying that the arguments are not presented "in order to persuade anyone of their truth, but only in order that they may provide a correct basis for calculation."[118] This made the ideas less potentially offensive.

Many astronomers discovered that Copernicus's theory made their calculations much easier, but they did not take it seriously as a literal description of the universe. Accordingly, the Church paid little attention to the theory.

In the 1590s, while teaching mathematics at the University of Padua, Galileo became convinced that the Copernican theory was correct, and he informally promoted it to friends and associates. As a result of his experiments and theories of motion, he had developed his concepts of inertia and the relativity of motion. He argued in a thought experiment that if a ship were sailing at a constant speed, then observers in an interior cabin would not be aware of this motion. This answered the main argument against the Copernican theory: we do not perceive the motion of the earth because we (and all of our surroundings) are moving with it.

Kepler's book *Mystery of the Universe* defended the Copernican system. When, in 1597, Galileo received a copy of this book, he wrote to Kepler:

> It is really pitiful that so few seek truth, but this is not the place to mourn over the miseries of our times. I shall read your book with special pleasure, because I have been an adherent of the Copernican system for many years. It explains to me the causes of many appearances of nature which are quite unintelligible within the commonly accepted hypothesis.[119]

Galileo's First Clash with the Church: 1616

IN 1609, GALILEO learned about a device invented in Holland that made distant objects appear closer. He quickly constructed his own improved design, and as he began directing his telescope at the heavens, he soon found new and dramatic evidence against the Ptolemaic system. First, he discovered that the moon was not the perfect sphere it was supposed to be as a heavenly body; it seemed to have a rugged surface similar to that of the earth.

The following year Galileo discovered four moons of Jupiter, proving that celestial bodies could orbit bodies other than the earth, which had not been thought possible. With the publication of these results in 1610 in *The Starry Messenger*, Galileo became famous throughout Europe.

Later that year, Galileo discovered that Venus had phases similar to the moon. These phases conclusively disproved the Ptolemaic system, and they provided strong evidence that Venus orbits the sun.

Copernicus held that Venus orbits the sun, but this was not unique to his system. In 1588 the Danish astronomer Tycho Brahe proposed a hybrid system, with the earth unmoving at the center of the universe, the sun and moon orbiting the earth, and the other planets all orbiting the sun. The Tychonic system became popular with astronomers in the early seventeenth century, because it had the computational advantages of Copernicus together with the religiously acceptable view that earth is fixed. The troublesome aspect of the Tychonic system was its bizarre asymmetry: virtually all the planets now orbited a moving point far from the center of the universe.

By the time Galileo discovered the phases of Venus, he had believed in the essential truth of the Copernican theory for over a dozen years, but he had only discussed it informally with friends. He had been cautious about promulgating the ideas in wider circles. But now, with the publication of *The Starry Messenger*, he was a widely-esteemed scientist, known throughout Europe. At this time, he had a much more secure position, being in the employ of the Grand Duke of Tuscany. He was ready to start openly advocating for his ideas.

Galileo wrote his first public defense of Copernicanism in 1613 as a result of a question asked of his student Benedetto Castelli. Castelli had been dining with the Grand Duchess Christina (mother of the Grand Duke of Florence). The Grand Duchess asked if the theory of Copernicus was consistent with the passage in the Book of Joshua, where God makes the sun stand still in order to lengthen the day. After Castelli told Galileo about this, Galileo wrote a long letter to the Grand Duchess. In this letter, which was meant for public dissemination, Galileo argued that the Bible should not always be taken literally; instead, we often need to interpret the Bible in the light of the natural knowledge we possess. In the story of Joshua, it is possible that God may have stopped the *earth's* motion, not the sun's. From the perspective of those who experienced it, the sun had stood still in the sky. Galileo argued that the Bible has no authority over natural science, which is based on reasoning from observations.

These comments were especially controversial in light of the Catholic Church's ongoing Counter-Reformation. The Council of Trent had resolved that the Bible could not be reinterpreted contrary to the views of the Church Fathers, who were undoubtedly not Copernicans.

On hearing about Galileo's letter, a number of academics and church-men smelled more than a whiff of heresy.

In late 1614, a Dominican Friar named Tommaso Caccini preached a sermon at Santa Maria Novella in Florence, in which he denounced Galileo in the strongest terms. A few months later, another Dominican, Niccolo Lorini, complained to the Inquisition.

Then a fellow Copernican jumped into the controversy. In early 1615, priest Paolo Antonio Foscarini published a book that interpreted the Bible to be compatible with Copernicus.

Hearing about the ongoing controversy, Galileo went to Rome in 1616 to argue his case. He wanted to try to persuade the Church to remain open on the question of whether the earth moves. A number of theologians and academics simply refused to look through the tele-scope, but others were more open to the new ideas. Galileo spoke to bishops, cardinals, and theologians, some of whom were sympathetic to the Copernican system.

A key figure in deciding the outcome of this issue was undoubtedly the Holy Inquisitor, Cardinal Roberto Bellarmino. Bellarmino was one of those who, in 1600, had condemned the Copernican Giordano Bruno to death for his unorthodox theological views. Bruno had been convicted of heresy and then burned alive. Bruno was not a scientist, but his cruel fate was a powerful reminder of the power of the Church to stifle views of which it did not approve.

On initially hearing about the telescope, Bellarmino had been inter-ested in learning more about science. But when he saw that the new views could cause theological problems, he lost any scientific curiously he might have had.

In February of 1616, the Holy Office of the Inquisition convened a panel of eleven experts to examine the Copernican thesis. They met at the Collegio Romano, where they deliberated for four days. On February 24 they concluded that the theory of Copernicus was formally heretical, with a vote of 11–0.

As a result of this recommendation, the Inquisition suspended Copernicus's book, placing it on the *Index of Prohibited Books* until it had been "corrected" by Church censors. Foscarini's book was condemned, and all copies were to be destroyed.

Galileo was summoned to an audience with the Holy Inquisitor. Cardinal Bellarmino ordered Galileo not to hold or defend the Copernican hypothesis in any way whatsoever. Galileo had no choice but to agree. Italy's greatest scientist had been silenced.

Since 1610 Galileo had been promising his supporters a book presenting his cosmological views, but now he was forbidden to work on it. The following years were the least productive of Galileo's career.

The early 1620s brought two events that gave Galileo hope for advancing his views. In 1621 Cardinal Bellarmino died. The Collegio Romano officially commemorated Bellarmino as "the hammer of the heretics." The engraving on his tomb summed up his life: "With force I have subdued the brains of the proud."[120]

In 1623, Galileo's old friend and admirer Cardinal Maffeo Barberini was elected pope, taking the title Urban VIII. Barberini had a reputation as an intellectual and had even written a poem for Galileo. Galileo saw a clear sign of hope that he could change the Church's position regarding Copernicus.

Meanwhile, Galileo had developed an argument for the earth's motion that he believed was conclusive. According to this argument, the earth's tides are caused by the moving of the earth. (It turns out that this is not a valid argument, but no one recognized this at the time.)

In 1624 Galileo made several visits to the new pope, and they discussed the possibility of the earth's motion. To Galileo's disappointment, the pope insisted that Galileo must not claim the Copernican system to be true. Galileo had no choice but to agree. However, the pope allowed that Galileo could discuss the theory as a mere hypothesis, providing that he included a certain argument. The argument was that since God is omnipotent, then He could have created the universe in any way He desired, so that any causal mechanisms we suppose to exist are merely hypotheses. Shortly after this meeting, the pope was heard to have said that the *Index*'s decree against Copernicus may have been a mistake.

Encouraged, Galileo returned to Florence and began work on his cosmological work, which he initially titled *On the Tides*, and later, *Dialogue on the Two Chief World Systems, Ptolemaic and Copernican.* (It did not directly address Tycho Brahe's system by name, but it included arguments that did address this system.)

Galileo's Second Clash with the Church: 1632–3

GALILEO SUBMITTED HIS dialogue to the censors of Rome in 1632, where it was approved, subject to a few modifications, and subsequently published. Written as an extended dialogue between three men, it was divided into four days of discussion. Galileo wrote it as not just a work of science, but also literature. In fact, it was a rhetorical masterpiece. This dialogue contained three participants: Simplicio, Sagredo, and Salviati. The name Simplicio was a play on the word for simpleton, and Simplicio was the dialogue's mouthpiece for the Church. In the dialogue, Salviati is the advocate of the Copernican system, Simplicio is the advocate of the Ptolemaic system, and Sagredo is the neutral character who asks questions, and is usually won over by Salviati. Simplicio, true to his name, is portrayed as a simple-minded fool.

Galileo did include the argument that the pope had insisted on, but it was on the last page of the book, in the mouth of the simpleton Simplicio. When this fact was pointed out to the pope, he was furious, telling the Florentine ambassador, "I have been deceived!" Galileo had not told the pope about Bellarmino's 1616 warning, and, further, had insulted the pope by putting his words in the mouth of a simpleton. In late 1632 the Holy Inquisition summoned Galileo to Rome.

Galileo had clearly disobeyed the order given him in 1616, but he was not about to confess to this. He argued that he was not presenting heliocentrism as true, but as merely a hypothesis. He admitted that he might have accidentally presented this hypothesis too strongly.

The legal case against Galileo focused on the question: Did Galileo violate the injunction of Cardinal Bellarmino? The Church's records on Bellarmino's meetings with Galileo had some irregularities, which worked in Galileo's favor. Another factor operating in Galileo's favor was the fact that the pope wanted the prestige of science; he did not want to appear anti-intellectual and opposed to new discoveries. According to physicist David Harriman,

> [T]he pope was caught in a contradiction. He wanted reason to function—but to remain subservient to faith. He wanted the products of independent thought—in order to honor and sanction the Church authority that opposed such thought. He wanted to claim the prestige of the new science—for the glory of the old religion. . . . In 1632, when the *Dialogue* was published, the Church authorities were forced to make a choice. If they wanted to be regarded as enlightened intellectuals, they would have to earn such a reputation by allowing the free exchange of ideas; if they wanted instead to uphold the long-standing rule of faith and force, then they must act openly as enemies of reason.[121]

Seeking a middle way out, the Inquisition settled on a less severe verdict than simple "heresy." They convicted Galileo of "vehement suspicion of heresy," which warranted a milder sentence. On June 22, 1633, in a crowded church, Galileo read the following statement:

> [A]fter having been judicially instructed with injunction by the Holy Office to abandon completely the false opinion that the sun is the center of the world and does not move and the earth is not the center of the world and moves, and not to hold, defend, or teach this false doctrine in any way whatever, orally or in writing; and after having been notified that this doctrine is contrary to Holy Scripture; I wrote and published a book in which I treat of this already condemned doctrine and adduce very effective reasons in its favor, without refuting them in any way; therefore, I have been judged vehemently suspected of heresy, namely of having held and believed that the sun is the center of the world and motionless and the earth is not the center and moves.

> Therefore, desiring to remove from the minds of Your Eminences
> and every faithful Christian this vehement suspicion, rightly con-
> ceived against me, with a sincere heart and unfeigned faith I abjure,
> curse, and detest the above-mentioned errors and heresies.[122]

Rumor has it that as the trial ended Galileo muttered "Nevertheless, it moves" under his breath. He probably did not actually say this, but it is clear that he was thinking it.

Galileo was sentenced to formal imprisonment for life, at the plea-sure of the Holy Inquisition. He was later released into the custody of Archbishop Ascanio Piccolomini in Siena, and the sentence was later commuted to perpetual house arrest at Galileo's small farm in Arcetri, near the convent where his daughters lived. He was forbidden to teach, travel, or even to visit his own daughters without permission. He became bitter about the injustice done to him "under the lying mask of religion" and he signed his letters "From my prison in Arcetri."[123]

The Church placed Galileo's dialogue on the *Index of Prohibited Books*, where it would remain for the next two centuries (until 1835). And even then, the Church still did not acknowledge any errors in its handling of Galileo.

Finally, in 1979, Pope John Paul II convened a commission to reinves-tigate the case of Galileo. The commission lasted thirteen years; then, in 1992, John Paul issued a formal announcement, saying in general terms that errors had been made, and that Galileo's interpretation of Scripture was actually the correct one. As a *Los Angeles Times* headline declared, "It's Official! The Earth Revolves Around the Sun, Even for the Vatican."[124]

Galileo and Modern Historians

WHILE MODERN HISTORIANS generally report the facts accurately about the historical events of Galileo's fall, they stress one or more of the following points:

- Using his powerful rhetorical skills, Galileo mercilessly ridiculed his enemies, who ended up reporting him to the Inquisition.

- The pope was struggling with political problems (such as his embroilment with the Thirty Years War), he was seen by many as weak, and he needed to demonstrate his assertiveness.
- The Church did not convict Galileo of "heresy" but only of the lesser charge "vehement suspicion of heresy."
- Although the inquisitors threatened Galileo with torture, he was never actually physically tortured or locked up in an actual prison.
- Galileo's primary proof of the earth's motion (based on the tides) is invalid.

What was the fundamental cause of Galileo's fall? There were many causes, answer the historians, and they are virtually impossible to rank. High on the list, they place Galileo's arrogance and the bad luck of the political situation. They do generally acknowledge that there was a conflict between science and religion here, but they deny the centrality of this conflict.[125] As one historian concludes, "Ultimately it may turn out that . . . underlying the apparent conflict between science and religion the trial of Galileo exhibits the deep structure of nothing less, and nothing more, than the conflict between conservatism and innovation."[126]

But the central conflict *was* between science and religion. Galileo may have been a sincere Catholic and had faith in many matters of religious doctrine, but when it came to the Copernican theory, he was fundamentally focused on evidence and logical reasoning. The Church, in censoring the arguments for the Copernican system, was not interested in the truth, but in maintaining its own authority. It is possible that if Galileo had played his cards differently—perhaps if he had put the pope's argument in the mouth of the neutral Sagredo—he might not have been summoned by the Inquisition, and he might not have been convicted of any crime. But this does not change the basic facts:

Galileo was fighting for a view he believed true based on evidence and reasoning, and the Church was blatantly suppressing debate on the subject in order to support a system based fundamentally on faith. Science was clashing with religion, in the most profound sense.

The Church widely disseminated its condemnation of Galileo. It then actively encouraged reactionary academics to attack the ideas of Copernicus, Kepler, and Galileo. As a result, numerous such critiques were published in Italy. One of these contained the following logical argument:

> Angels make Saturn, Jupiter, and the sun turn around. If the earth revolves, it must also have an angel in the centre to set it in motion. But only devils live here, so it would therefore be a devil who would input motion to the earth. . . . [I]t seems, therefore, to be a grievous wrong to place the earth, which is a sink of impurity, among the heavenly bodies which are pure and divine things.[127]

It was now clear to scientists that if they wanted to discuss the new theories of the universe openly, they needed to stay clear of the Catholic Church. The French philosopher René Descartes had been writing a treatise titled *The World*, which made use of Copernican principles, but on hearing of Galileo's fate, Descartes withdrew his publication plans. During the Renaissance, Italy had been the center of gravity of the intellectual life of Europe, but now this center shifted northward, and never returned to Italy.

* * *

Galileo's conflict with the Church was indeed a conflict between religion and science. Galileo's reliance on evidence and reasoning was implicitly based on the *primacy of existence*. The Church's reliance on authority and faith was based on the *primacy of consciousness*: the consciousness of the Church authorities, and ultimately, that of God.

CHAPTER EIGHT

The Scientific Revolution

T HE SCIENTIFIC REVOLUTION of the sixteenth and seventeenth centuries was a defining moment in the history of Western Civilization. Modern science and the scientific method were born; the rate of scientific discovery exploded; giants such as Copernicus, Vesalius, Kepler, Galileo, Harvey, and Newton, as well as countless lesser figures, unlocked world-changing secrets of the universe.

It has often been observed that such a revolution occurred only once in human history and in one particular culture: the predominantly Christian culture of early-modern Europe. This observation gives rise to several questions: What role, if any, did Christianity play in the birth of modern science? Did faith give rise to science? Did a mixture of faith and reason give rise to it? Was Christianity somehow responsible—or even *necessary*—for the rise of modern science, as some historians have argued? In short, what, if anything, does God have to do with the Scientific Revolution?

As long as science has existed, religionists have been attempting to reconcile religion and science. But recently, a new breed of scholars has asserted that religion itself—Christianity in particular—actually caused the birth of science. What are the facts of the matter?

The bulk of this chapter will deal with these issues. Before turning to them, we will first take an overview of the main developments of the Scientific Revolution. The following sections will examine several significant changes—integral to the Scientific Revolution—that were occurring regarding the way people viewed science and religion.

The Scientific Revolution: Key Developments

THE YEAR 1543 saw the publication of two monumentally important scientific works, and for that reason it is often considered the beginning of the Scientific Revolution. The first was Andreas Vesalius's *On the Structure of the Human Body*, which, with its astounding illustrations, founded the modern study of anatomy. Vesalius, a trained physician, was deeply influenced by the naturalism of the Renaissance—by the idea that the natural world merits our careful study. This led him to perform meticulous dissections of human cadavers and to hire an accomplished Renaissance artist to create the fabulously-detailed woodcut illustrations for his book. The work of Vesalius inspired many followers who later made important anatomical and physiological discoveries.

But even more significant than Vesalius's book, as we've seen in the previous chapter, was Nicolaus Copernicus's *On the Revolutions of the Heavenly Spheres*, published that same year. The heliocentric system of Copernicus formed the first link in a chain of astronomical observations, reasoning, and discoveries that included those of such great scientists as Tycho Brahe, Johannes Kepler, and Galileo Galilei, and which culminated in a grand synthesis on the part of Isaac Newton, with his universal law of gravitation. This chain of discoveries would not have been possible without Galileo's brilliant experimental investigations into the nature of free fall using pendulums and inclined planes. Galileo's contemporaries William Gilbert and Francis Bacon also focused on the idea of systematic experimentation as the key to understanding natural phenomena. As the seventeenth century progressed, belief in the power of experimentation became widespread, and scientists invented new precision tools for performing observations and experiments. These instruments included the telescope, microscope, barometer, thermometer, precision clock, and vacuum pump.

Galileo was the first of many brilliant experimentalists who found ingenious ways of investigating the properties of nature. One of these men, Robert Boyle, used the barometer and vacuum pump to discover the exact relationship between the volume of gas and the pressure exerted on it (known today as Boyle's law). He also subjected alchemical ideas to systematic experimentation and rational analysis, and is thus considered one of the forefathers of the science of chemistry.

These new scientists were eager to share their ideas with like-minded others, and soon conceived of an entirely new type of social organization: the scientific society. Scientific societies in Italy, France, England, and elsewhere made it much easier for scientists to meet and discuss their work, to perform experiments, and to report on scientific work done in other countries. The earliest scientific society, to which Galileo belonged, was the Accademia dei Lincei (Academy of the Lynx) in Rome, which was succeeded by the more formal Accademia del Cimento (Academy of Experiment). By the 1660s, royally-sponsored societies had been founded in France (Academie Royale des Sciences) and England (The Royal Society of London), and both had official journals for the publication of research done by their members.

The sciences most fundamental to the Scientific Revolution were astronomy and physics, but scientists also made significant discoveries regarding the nature of living things: that blood circulates through the body; that reproduction begins with the union of egg and sperm; that sexual reproduction occurs in plants; and that life functions could be viewed as analogs of physical and chemical processes.

The most important discovery of the Scientific Revolution was the powerful new method of discovery itself. The modern scientific method relied on a combination of careful observation, controlled experiments, and the search for mathematical laws and their relationships. Historian of science I. B. Cohen points out

> how novel and revolutionary it was to discover principles by experiment combined with mathematical analysis, to set scientific laws in the context of experience, and to test the validity of knowledge by making an experimental test. Traditionally, knowledge had been based on faith and insight, on reason [detached from observation] and revelation. The new science discarded all

of these as ways of understanding nature and set up experience—experiment and critical observation—as the foundation and ultimate test of knowledge.[128]

This new method was ultimately an extension of Aristotle's basic philosophical orientation: his this-worldly metaphysics and his epistemology of reason.

Unfortunately, Aristotle's specific theories about physics and cosmology, such as his explanation of why heavy objects fall, proved to be incorrect. Also, a number of Scholastic philosophers insisted on the truth of these Aristotelian theories even in the face of new observational evidence against them. As a result, many seventeenth-century scientists formed a highly negative judgment of Aristotle, even as they accepted his basic philosophical orientation. As scientists such as Galileo and William Harvey pointed out, Aristotle had argued that all science must be tied to careful observation of the world. Philosophically, Aristotle's *primacy of existence* was the foundation of the Scientific Revolution. It was owing to Thomas Aquinas that Aristotle's ideas were back on the scene.

Making Room for Reason: God's Two Books

DURING THE SCIENTIFIC Revolution there was little explicit conflict between science and religion, apart from Galileo's famous clash with the Roman Inquisition. Scientists generally adopted (or at least did not openly reject) the religious views of their culture. However, there was a significant reorientation between science and religion, in which people reconceived God as playing a much smaller role in the universe. One key to this reorientation was the metaphor of *God's Two Books*—the Book of Scripture and the Book of Nature—which were considered equally deserving of our attention. This metaphor originated in Thomas Aquinas's distinction between the realms of faith and reason, and was famously promoted in the seventeenth century by Francis Bacon:

> Let no man upon a weak conceit of sobriety or an ill-applied moderation think or maintain, that a man can search too far, or

be too well studied in the book of God's word, or in the book of God's works, divinity or philosophy; but rather let men endeavour an endless progress or proficience in both; only let men beware . . . that they do not unwisely mingle or confound these learnings together.[129]

Bacon's warning about mingling the two books became highly influential among his many seventeenth-century followers, and it became widely accepted that God's two books should be studied quite separately. As far as religious belief was concerned, Bacon urged men to "give to faith only that which is faith's."[130] This attitude was liberating for scientific work, keeping it largely unmolested by religion.

The opposite attitude—that religion is closely connected with the study of the physical world—had led to the Church's persecution of Galileo. He had advocated an idea—that the Earth moves—that conflicted with Biblical passages implying that the Earth does not move.

Bacon's writings were influential throughout the new scientific societies in Europe, but they were especially so in England, where the founders of the Royal Society of London referenced them explicitly. As a result of this influence, at the Royal Society, "no one ever presented a public case for a scientific fact with a theological argument."[131]

Historians have often pointed out that many of the seventeenth-century English scientists had careers in the Church. But even these "theologian-scientists" sought to isolate their religious beliefs from their scientific studies, in effect leading "double lives": "English scientists *qua* scientists kept out of the sacristy, English theologians *qua* theologians kept out of the rooms where experiments were performed."[132]

Among the accomplished scientists, the two figures most famous for their strong religious beliefs were Robert Boyle and Isaac Newton. But even these two were careful to keep a separation between God's two books. According to one historian, "When working as a 'naturalist,' Boyle sought to 'discourse of natural things' only, without 'intermeddling with supernatural mysteries.'"[133] Newton wrote extensively on religion, but his scientific books contained no religious arguments for his scientific conclusions. And when he was president of the Royal Society, "he banned anything remotely touching on religion."[134]

The Clockwork Universe

AROUND THE YEAR 1600, when astronomers Tycho Brahe and Johannes Kepler came to Prague to accept employment from Holy Roman Emperor Rudolf II, they could not have failed to notice Prague's great astronomical clock. Mechanical clocks had started appearing in Europe in the late thirteenth century, and had become widespread in larger cites by the mid-fourteenth century. The first clocks had their innards exposed, so everyone could easily see how they worked. By the sixteenth century, clocks were generally built to conceal their inner workings. Many clocks, such as the ones in Prague and Strasbourg, did much more than just tell time; they indicated solar and lunar cycles, and marked key moments with startling animations such as crowing roosters and jousting knights.

The inside of a clock involved extremely complex motions that could even appear purposeful. This led some to start thinking that perhaps the entire cosmos was like this in some way. During the seventeenth century, people began seeing nature as essentially similar to a machine, and the clock became a favorite metaphor. In reaction to the natural magic tradition of the Renaissance, with its "sympathies" and "antipathies" and "occult" powers, the new view considered nature as a machine made up of unconscious material parts controlled by external mechanical forces. In its extreme form, this led to the idea that all the phenomena of nature are produced by particles of matter in motion. When Robert Hooke looked at insects under the microscope, he was amazed at the intricate details he saw; these made him think of the insects as complex machines. Robert Boyle, one of the greatest champions of this new way of looking at the world, called it the "mechanical philosophy."

The adherents of the mechanical philosophy had a variety of opinions about the nature of matter and mechanical forces. Catholic priest Pierre Gassendi followed the atomist tradition from Democritus of Ancient Greece, which held that all matter is composed of indivisible solid particles. Others such as René Descartes believed that matter is infinitely divisible. Some saw matter as entirely inert, and others saw it as having some active capacities. While Newton's law of gravity operated without direct physical contact between objects, and was disputed by some for that reason, it was seen by others as still essentially keeping to the mechanical tradition.

From its first formulations, the mechanical philosophy invited charges of atheism because it suggested that the world machine could run itself without the external help of a God. In defense of their piety, its proponents always stressed that they saw God as the creator and constant operator of this machine. Invoking the distinction between "primary" and "secondary" causes as the Mu'tazilites had done centuries earlier, they explained that God was, of course, the "primary" cause of all events and that they were studying "secondary" mechanical causes. In this way, they gradually came to see God less and less as an explanation for anything.

As the culmination of the Scientific Revolution, Newton's *Mathematical Principles of Natural Philosophy* demonstrated that the world works on essentially mechanical principles understandable by human beings. Although for Newton, God was still there actively adjusting and supporting the universe in its operations, many of his successors held that once the universe was up and running it shouldn't require outside intervention.

Natural Theology Displaces Revealed Theology

When Thomas Aquinas claimed that there were theological truths that could be established by reason alone, he created the field of "natural theology." Aquinas argued that reason (and thus natural theology) could prove the existence of God and several of God's characteristics. For the rest of the essential truths of Christianity, one needed to turn to revealed theology: the domain of faith, revelations, and the Bible. Revealed theology was necessary to establish the central doctrines of Christianity, such as the Trinity and Christ's role as redeemer for our sins.

The study of natural theology became popular in the seventeenth century, especially among English theologians. Many of the seventeenth-century scientists were excited about the discoveries they were making, and they associated this excitement and wonder with religious feeling. Robert Hooke expressed a common view:

> 'Tis the contemplation of the wonderful order, law, and power of that we call nature that does most magnify the beauty and excellency of the divine providence, which has so disposed, ordered, adapted, and empowered each part so to operate as to produce the wonderful effects which we see; I say wonderful because every natural production may be truly said to be a wonder or miracle if duly considered.[135]

Another Royal Society member spoke about "the lawful and religious delight which should result from beholding the curious and wonderful frame of this our visible world."[136] By the year 1700, natural theology had become so popular that it displaced much of revealed theology.

Robert Boyle—whose religiosity has already been noted—was one of the greatest champions of natural theology among the accomplished scientists of the era. He wrote voluminously about how science proved the greatness of God and how scientific research was a form of worship, and his book *The Christian Virtuoso* (where "virtuoso" meant "scientist") described how doing science makes a man a better Christian. Although Boyle was no doubt sincere in his profession of religious belief, historian Richard Westfall has noted a major omission in Boyle's writings. For all his professed devotion to Christianity, Boyle barely mentioned Christ and essentially ignored the core doctrine of his religion:

> The scarcity of references to Christ in the many pages of Boyle's disquisitions on religion is striking. . . . The idea of redemption did not play an important role in his thought on moral living; he considered the attainment of virtue as an intellectual choice dependent on a person's comprehension of God. . . . The problems of morality were not pressing concerns in his personal life. He talked about sin like an American discussing cricket; he had

heard about it but had never seen it close at hand. The Christian
doctrine of redemption rang no response in his soul.[137]

The shift of intellectual focus from revealed theology to natural theology was a step away from the Christian God and toward the deist God,
which in turn was only one step away from a universe devoid of the
supernatural.

Bearing in mind the key developments leading up to and during the
Scientific Revolution, the prevailing religious attitudes of its scientists,
and the gradual waning of God as an explanation for nature, we can
return to a question raised at the outset of this chapter.

Was Christianity Necessary for the
Birth of Modern Science?

IN RECENT DECADES, academic historians of science have rejected the
idea that science and religion are incompatible, and they have written
volumes in an effort to refute what they call the "conflict thesis"—the
idea that science and religion are fundamentally in conflict. (See the
Appendix for the history of the "conflict thesis.") These historians
have also acted as enablers for diehard religionists to argue that religion caused the birth of science. These religionists include the prolific
Stanley Jaki, Benedictine priest and professor of physics, Ian Barbour,
professor of religion and physics, Thomas Woods, a popular libertar-
ian Catholic writer, and Rodney Stark, a sociologist of religion at Baylor
University. Stark, the most prominent contemporary proponent of this
view, devotes a chapter of his book *For the Glory of God* to the idea that
Christianity laid the philosophical groundwork for the development
of modern science, writing, "Christianity depicted God as a rational,
responsive, dependable, and omnipotent being and the universe as his
personal creation, thus having a rational, lawful, stable structure, await-
ing human comprehension."[138]

Although many of the seventeenth-century scientists had such a conception of God and the universe, this is no ground for attributing the

rise of science to Christianity. Scientists were not studying the "super-natural realm"; they were studying observable nature, the identity of things, and causal relationships. They were employing not faith but reason, observation, logic. Had scientists tried to ground science in religion, they would have been utterly stifled. Where would they have turned? To God? He can't be observed; this is both a tenet of religion and a simple fact. The only place scientists could have turned to "ground" science in religion is the Bible. What would they have found there?

The Bible certainly does not present God as "rational, responsive, and dependable" or the universe as "having a rational, lawful, stable structure" that is open to "human comprehension." The Old Testament presents the natural world as created by a supernatural being who frequently acts to intervene in his creation, to make things act in contradiction to their natures—by, for instance, stopping the sun and moon from moving, turning a woman into salt, and making a bush speak. Likewise, the New Testament presents a world full of supernatural, unscientific, causally impossible events—from Mary being impregnated by a ghost, to Jesus walking on water, to a dead man rising from the grave. (And this is to say nothing of the moral atrocities perpetrated by God in the Bible—from His exterminating most of the human race in a flood to his torturing Job.) The whole Christian worldview entails the subordination of reality, identity, and causality to the whims of an alleged God for whom there is no evidence and who is therefore to be accepted on faith.

Why then, did seventeenth-century scientists think of God as "rational" if they could not have gotten this idea from the Bible? The conception of a "rational" God, a God that somehow fit with nature, was a projection of their *own* minds—minds that were beginning to embrace a rational view of the world, and thus turning away from the philosophical and literary core of Christianity. After Aquinas introduced Aristotelian principles to the Christian world, Christian culture acted as a carrier of the pro-reason, pro-reality philosophy. Aristotle's philosophy was always at odds with the supernatural and mystical core of Christianity, which is why Aquinas himself had to divide knowledge into two separate realms. In spite of its historical longevity, Aquinas's theology was always an unstable amalgam of contradictory components, one of which is the foundation of science, the other of which contradicts it.

Modern science was born in a culture in which Christianity was the dominant religion, but modern science was not built on a Christian foundation. It was built on the implicit foundation provided by the pro-reason, *primacy-of-existence* philosophy of Aristotle.

* * *

The great discoveries of the Scientific Revolution demonstrated the awesome power of the rational human mind. Since the fall of Rome, Christianity had dominated all cultural and intellectual activity in the West. But by the end of the seventeenth century, Christianity had lost its dominance in the realm of the intellect, and science had become the primary source of inspiration for those eager to use their minds.

Although explicit religious belief remained at high levels among the new scientists, it moved from the foreground to the background of their minds. Aquinas had segregated religion from reason, enabling reason to flourish. Revealed theology—true religion—was displaced by natural theology, dramatically loosening the shackles of religion and inspiring people to observe reality and think. The universe was increasingly seen as a machine that God had designed and then left alone, enabling us— through reason—to discover the laws of nature and to transform the world to suit our needs. A vestigial God remained as an "explanation" of the universe, but future generations would realize that God was not needed as an explanation for anything at all.

The claim of today's religious apologists that the Scientific Revolution is rooted in religion—that science somehow came from faith—is historically and logically absurd. The foundation and cause of the Scientific Revolution was not religion or faith but observation and logic.

In Chapter Five, we saw how the Renaissance caused the *primacy of consciousness* to lose strength. The Scientific Revolution entailed a further shift in mental attitude away from the *primacy of consciousness*, and toward the *primacy of existence*.

CHAPTER NINE
Newton, Rationality, and the Arbitrary

T HE GREATEST SCIENTIST of the Scientific Revolution was undoubtedly Isaac Newton. His spectacular discoveries in optics, mechanics, and astronomy created a foundation for virtually all of the science of the following centuries.

We will see in the next chapter that the impact of Newton's discoveries extended far beyond science, and led to a society-wide confidence in reason that would characterize an age. English poet Alexander Pope wrote Newton's epitaph: "Nature and Nature's laws lay hid in night; God said, let Newton be, and all was light."

In his scientific work, Isaac Newton was fundamentally a practitioner of *inductive* logic, which starts with observations of reality and integrates them into wider and wider generalizations. (The other type of inference—deductive logic—applies general truths such as "all mammals bear live young" to more specific instances, such as zebras, allowing one to conclude, for example, that "since zebras are mammals, they must bear live young.")

As Newton explained:

> Natural philosophy consists in discovering the frame and operations of nature, and reducing them, as far as may be, to general rules or laws—establishing these rules or laws by observations and experiments, and thence deducing [i.e. inferring] the causes and effects of things . . .[139]

As we will see, Newton identified a crucial principle of inductive logic—a principle that, when applied consistently, fatally undercuts religious ideas.

The Discovery of the Laws of Motion

As NEWTON HIMSELF pointed out, he "stood on the shoulders of giants." Two of these giants were Galileo and Kepler, who made key discoveries in mechanics and astronomy.

Galileo's observations and experiments had led him to his law of falling bodies (with constant acceleration) and a simple concept of inertia. Kepler's study of the orbit of Mars had led to his three laws of planetary motion:

1. The shape of a planet's orbit is an ellipse with the sun located at one focus.
2. A planet sweeps out equal areas in equal times.
3. The squares of the periods of revolution of any two planets around the sun are proportional to the cubes of their mean distances from the sun.

Starting with these laws, Newton developed a new concept of "force" and three laws of motion, which he lays out in the first section of his *Mathematical Principles of Natural Philosophy*, published in 1687:[140]

1. Every body perseveres in its state of being at rest or of moving uniformly straight forward, except insofar as it is compelled to change its state by forces impressed.
2. A change in motion is proportional to the motive force impressed and takes place along the straight line in which that force is impressed.
3. To any action there is always an opposite and equal reaction; in other words, the actions of two bodies upon each other are always equal and always opposite in direction.[141]

Newton's most brilliant insight was to see that a single force does all of the following:

- keeps planets in their orbits around the sun
- keeps moons (such as our moon and Jupiter's moons) in their orbits
- causes objects to fall
- holds objects on the earth
- causes the tides

Newton organized his *Principia* in a rigorously deductive fashion, with the three laws of motion presented like mathematical axioms, from which he seemingly deduces everything else, one after another. But this organization (following the convention of other scientific works of the time) is misleading; it is essentially the reverse of the order in which he reasoned.

Newton's Rejection of "Hypothesis"

THE SECOND EDITION of Newton's *Principia* included a "*General Scholium*"—a general note at the end—where he famously said "Hypotheses non fingo" ("I do not feign hypotheses.")[142] His immediate point here was that he was not proposing a deeper mechanism that caused and explained the law of gravity, as Descartes had attempted to do. Newton was aware that there was no evidence for a specific mechanism, and that the mechanism proposed by Descartes contradicted known facts. In his note, Newton pointed out that he had discovered an important law of nature, and that this should be enough.[143]

Newton's statement that "I do not feign hypotheses" was also making a broader point, which can be seen in his approach to the study of light and color. This is the subject of his other major publication: *Optics, or, A Treatise of the Reflections, Refractions, Inflections & Colours of Light* (first published in 1704).

After the use of telescopes became widespread, their users discovered that telescopic observations revealed bits of colorings that made the image fuzzy; today these are referred to as *chromatic aberrations*. People also knew that rainbows could appear when light went through raindrops, and that colors result from light going through prisms. Somehow, it was understood, that when pure white light was refracted, i.e., bent, at an interface (such as between glass and air), it was modified in some way to generate colors.

Newton studied the theories of light and colors from Descartes, Hooke, and Boyle, but he found these unconvincing and essentially based on imagination. Deciding to undertake his own research of the subject, he purchased a prism and began experimenting with it in a darkened room with sunlight coming in through a pinhole. Surprisingly, the light, when refracted through the prism and shown on the wall, did not make a circular shape; instead, it made an elongated shape with straight sides and circular ends (similar to an ellipse). One end had a red fringe; the opposite end had a blue fringe.

Newton started wondering if blue and red were somehow already in the white light, but were refracted at different angles, so he tried an experiment. He took a straight thread and colored half of it blue, and the other half red. Then when looking at the straight thread through a prism, he saw the two halves as discontinuous and not in a straight line. This was a powerful piece of evidence that he was on the right path.

This led him to a series of further experiments with the prism and other objects, including the famous experiment with two prisms, in which he separated white light into colors and then reconstructed it. This clearly demonstrated light's heterogeneity. Newton concluded that white light consists of a mixture of all colors, which are separated (but not otherwise modified) by refraction.

Newton's approach to studying the nature of light differed in a key way from the approaches of contemporaries such as Descartes and Hooke, who began with certain assumptions about the nature of light, and deduced the consequences of these. Descartes speculated that light consists of rotating particles with the speed of rotation determining the color. Robert Hooke speculated that white light consists of a symmetrical wave pulse, which results in colors when the wave becomes distorted. But these speculations had no real evidence in their favor, and these were the type of "hypotheses" that Newton rejected.

In a letter to a friend, Newton explained his attitude toward these "hypotheses":

> [I]f anyone offers conjectures about the truth of things from the mere possibility of hypotheses, I do not see by what stipulation anything certain can be determined in any science; since one or another set of hypotheses may always be devised which will appear to supply new difficulties. Hence I judged that one should abstain from contemplating hypotheses, as from improper argumentation . . .[144]

This was Newton's rejection of what Rand would call the "arbitrary." Newton identified an important epistemological principle: The arbitrary is not to be entertained as possible; otherwise, all our knowledge is invalidated:

> whatever is not deduced [i.e. inferred] from the phenomena must be called a hypothesis; and hypotheses, whether metaphysical or physical, or based on occult qualities, or mechanical, have no place in experimental philosophy. In this experimental philosophy, propositions are deduced [i.e. inferred] from the phenomena and are made general by induction.[145]

An arbitrary idea is one for which there is no evidence. Like the dragon in Carl Sagan's garage, it is a notion put forth based solely on whim or faith.[146] Rand held that an arbitrary idea cannot be valid even as a possibility; in order to say "it is possible," one needs to have evidence (which can consist of either direct observations or reasoning based on observations).

This rejection of the arbitrary may be expressed in a positive form: One should be focused on reality, and only on reality.

The Arbitrary as Non-Cognitive

AN ESSENTIAL PRINCIPLE of rational thinking is what some logicians have called the Law of Rationality,[147] and which philosopher Harry Binswanger sums up as follows: "In reaching conclusions, consider all the evidence and only the evidence." He elaborates:

> Because evidence is the only means of gaining inferential knowl-
> edge, the rational mind accepts all that which the evidence
> shows, only that which the evidence shows, and only to the extent
> that it shows it. Only evidence — not someone's assertions, not
> feelings, not authority, not faith — can provide the basis for pro-
> ceeding cognitively.[148]

Ideas not based on evidence must be classified as *arbitrary*, which is an epistemological status different from *true* or *false*.[149] Consider a simple example: Suppose I am getting ready to leave my house, and I can't find my keys; they are not on the bed stand where I usually leave them. Suppose the thought then strikes me that a monkey has taken my keys. (I watched a documentary about African primates last night.) I have no evidence of any monkey ever being in or near my home. So, this idea is arbitrary, and the only rational course for me is to dismiss it.

Of course, it is not physically impossible for a monkey to take my keys; however, for me to claim, "it is *possible* that a monkey took my keys" would be unwarranted and arbitrary. It is solely based on my imagina-tion. Imagination is useful in many ways (such as in entertainment), but it is not cognition (awareness of reality), and it should never be con-fused with cognition.

To say "maybe" or "it is possible" imposes an obligation on the per-son who says it—the obligation to provide some evidence in support of the claim.[150] In today's culture, one often hears someone advance an arbitrary claim, and then say "prove it isn't so." But there is no burden of proof on the person who dismisses the arbitrary. In fact, it is an error even to attempt to disprove an arbitrary claim.

Sometimes one can relate an arbitrary idea to a cognitive context, and then show that the idea contradicts existing knowledge. Suppose that someone suggests—without any evidence—that there is a zebra grazing in my backyard. Given that this assertion is arbitrary, I should simply dismiss it. But suppose I go ahead and look, and I see that the yard is empty. Then I can conclude that there is no zebra there; then the claim becomes *false* instead of *arbitrary*. The advocate of the arbitrary can then go ahead and insulate his claim from my observation; he can say "the zebra in your backyard is invisible and undetectable. Prove that my claim is false." The only rational response to this is: "Your claim is arbitrary; therefore, I refuse to consider it."

This discussion has merely touched on some of the issues regarding the role of evidence in reasoning and the invalidity of the arbitrary. For a detailed discussion of these issues, see *How We Know: Epistemology on an Objectivist Foundation* by Harry Binswanger.

Unfortunately, in today's world, arbitrary ideas are routinely considered in a variety of fields, including philosophy and law.[151] The most prevalent source of arbitrary ideas is emotion, so let us examine the nature of emotions.

The Nature of Emotions and Faith

THE UNCOMPROMISING ADVOCATE of reason is often met by the objection: "But what about emotion?" Emotions have often been taken to give us some special knowledge, by a special means, other than reason.

But this is not what they are. An emotion is an automatic, psychosomatic response to a situation, a response that reflects one's understanding and evaluation. Emotions merely involve the application of our existing ideas and evaluations—right or wrong—to the situation. They occur so quickly and automatically that it is easy to escape this fact, but it is crucial never to forget that emotions are a product of ideas—which can be wrong.

Every emotion we experience involves a chain of events containing four elements:

1.   perception (I see a curved shape on the ground.)

2. identification (That is a rattlesnake.)
3. evaluation (My life is threatened.)
4. reaction (I feel fear.)

This chain occurs virtually instantaneously, and we are typically aware only of the first and last elements: perception and reaction. I see a snake and I feel fear.

People often experience conflicts between their reason and their emotions, but these are more accurately described as conflicts between their accepted ideas. For example, a man with an irrational fear of flying may tell himself that he recognizes that flying is safe, but deep down, he doesn't really believe this. Part of the process of overcoming this problem involves fully integrating his thoughts and conclusions.

Emotions do not have independent access to reality. The only way to know if an emotion is appropriate is by examining it rationally.

This is not an anti-emotion viewpoint. Emotions play crucial roles in our lives; for example, as sources of enjoyment, motivation, and self-awareness. But they do not directly tell us about external reality. As Rand put it, "emotions are not tools of cognition."[52]

Faith amounts to the acceptance of an idea based on emotion rather than of evidence. It is believing because you want to believe: "I want to believe in Jesus and that he loves me; therefore he does." Faith undercuts the mind's need for evidence. The acceptance of faith leads to the complete rejection of the Law of Rationality. This is why Rand claimed, "The alleged short-cut to knowledge, which is faith, is only a short-circuit destroying the mind."[53]

The person who accepts faith implicitly accepts the *primacy of consciousness*. To say, "I want X to be true; therefore it is true" is to say, "my consciousness can directly control existence."

If the existence of God could be rationally proved, then it would not be arbitrary. Theologians such as Thomas Aquinas thought they had such rational proofs. But, as many have pointed out, these are all fatally flawed. (For a good discussion of these alleged proofs, see *Atheism: The Case Against God*, by George H. Smith.)

The main reason people believe in God is not any rational argument, but their feelings. God is a father figure who makes people feel reassured that someone loves them, and will take care of them. And the popular argument, "If there is no God, who created the universe?" clearly presupposes the *primacy of consciousness*.

Newton's Inconsistency

UNFORTUNATELY, NEWTON DID not see the full implications of his refusal to "feign hypotheses." When it came to his religious beliefs, he accepted most of the conventional views of the English Puritan culture in which he was raised. He did not ask for proofs or evidence when it came to religion.

Newton trusted the Bible as a historical document that was accurate except for certain "corruptions." He often took the Bible literally, especially prophetic texts from Daniel and Revelation. He believed in predestination, the bodily resurrection of Jesus, the future resurrection of the faithful, and the millennial kingdom ruled by Christ. He even calculated a possible starting date for the events of the Apocalypse: the year A.D. 2060. His private manuscripts on theology are voluminous, totaling four million words.

While it is unfortunate that Newton spent so much of his time and intellectual energy on such pointless pursuits, it is important to remember him for his inspiring achievements, and the extent to which he did— with spectacular results—apply his logical reasoning to science.

CHAPTER TEN

The Discovery of Deep Time Leads to the Demotion of Genesis

T HE ACHIEVEMENTS OF the Scientific Revolution proved the power of the rational human mind, which was now confident enough to challenge established religious dogmas. This intellectual transformation of Europe led to the Enlightenment—an era marked by a supreme confidence in the power of the rational mind together with an antipathy for superstition and dogmatism. It became increasingly acceptable to criticize faith-based Christian dogmas such as the Trinity or the role of the sacraments. Many thinkers turned to deism—the view that God created the world and then withdrew from taking an active part in it. Deists accepted the Bible's moral authority, but rejected its literal interpretation as an accurate history of the world; God had created the universal laws governing the universe, but He certainly did not tinker day to day in the world, through events such as miracles and floods. The deists saw the natural world not as a ruin of a more perfect time, but as an orderly world that is open to human achievement. In addition to the deists, there were even a few self-proclaimed atheists, such as the Baron d'Holbach and Denis Diderot (creators of the famous *Encyclopedie*).

Enlightenment thinkers saw that modern science relied on a view of nature as being completely causal, and that the existence of miracles was a direct contradiction to this view of nature. This confidence in science led them to have scorn for the Biblical stories of miracles and to see a fundamental conflict between science and religion. The Marquis de Condorcet wrote that Christianity feared that spirit of doubt and inquiry, that confidence in one's own reason, which is the bane of all religious beliefs. The natural sciences were odious and suspect, for they are very dangerous to the success of miracles; and there is no religion that does not force its devotees to swallow a few physical absurdities.[154]

The momentum of scientific discovery from the seventeenth century continued into the eighteenth. One relatively new science was geology. Observations of the earth's surface often revealed rock layers containing shapes of shells and fish; these layers could have been sedimentary deposits at the bottoms of ancient seas. The shapes embedded in the rocks led to questions about how they got there, especially the animal shapes which did not correspond to any existing animals. Earthquakes and volcanic eruptions had always prompted people to wonder about their causes; during the Middle Ages, many saw these as direct signs of God's anger, but now they considered possible natural causes. Numerous thinkers worked on reconciling these observations with Biblical events such as the Flood. As more and more of these observations were made, it started becoming obvious that the Earth was much older than the 6000 or so years that the Bible seemed to indicate.

Some of the figures of the Scientific Revolution started making observations and starting a chain of reasoning that would eventually lead to the embrace of a very different timescale. The new timescale would truly dwarf the Biblical account, since it was measured in tens of millions or even billions of years.

By the time Charles Darwin started conceiving his theory of natural selection in 1838, he could be confident of the immensity of time that had elapsed in the earth's history—an immensity of time that was a requirement for his theory of evolution. The idea of a timescale so immensely removed from our ability to grasp perceptually has been called *deep time*, and it is central to the science of geology.

Archbishop James Ussher and the Age of the Earth

THE BIBLE PROVIDES a narrative of interconnected events, starting with the creation of the world by God in Genesis 1:

> In the beginning God created the heavens and the earth. The earth was without form and void, and darkness was upon the face of the deep; and the Spirit of God was moving over the face of the waters. And God said, "Let there be light"; and there was light. And God saw that the light was good; and God separated the light from the darkness. God called the light Day, and the darkness he called Night. And there was evening and there was morning, one day.[155]

The following verses describe the creation of the "heavens" on the second day; the dry land, the seas, and vegetation on the third day; the sun, moon, and stars on the fourth day; the birds and the fish on the fifth day; and finally, animals and man on the sixth day.

Ever since the Patristic period of Christianity, biblical scholars have performed calculations based on the Bible and Jewish historical traditions to assign dates to various biblical events. In the fourth century, Eusebius of Caesarea assembled a chronicle based on Jewish historical traditions reaching back in time to the birth of Abraham. Saint Jerome extended this chronicle back to Adam.

The Protestant Reformation in Europe led to a proliferation of Biblical chronologies. The most definitive and well-known calculation of the age of the world was performed by the Irish scholar and Anglican Archbishop James Ussher (1581–1656). Using the King James Bible as his source, Ussher added together all the ages and reigns of the kings, and concluded that the world was created in 4004 B.C. Further information, including astronomical tables, led him to conclude that the world was created in the evening preceding October 23 of the year 4004 B.C. Ussher published his conclusion in his *Annals of the World* in 1650. His dates were reprinted in the margins of numerous editions of the King James Bible starting in 1701.

Many Christians considered the chronology of the world a crucial subject, because they considered it a clue as to the date of the end of the world and the Last Judgment. Many believed that the world would last 6000 years in total.[156]

Nicolaus Steno and the Nature of Fossils

SINCE ANTIQUITY, PEOPLE had noted that the shapes of shells and fish could sometimes be found in rock layers which were far from the sea, and sometimes even high up in the mountains. Two competing theories had emerged as to their origins. One was that actual fish had somehow become petrified. The other view was that the fish shapes had gotten imprinted on the rocks as accidents or "sports of nature;" the platonic Form of "fish" that was intended to get into the sea had accidentally ended up in rock, causing the fish-shape to grow in the rock.

The first significant steps toward a rational view of the earth's past were taken by Danish physician Nils Stenson, better known as Nicolaus Steno (1638–86). Steno attended the University of Copenhagen, where he studied anatomy. His interest in rocks came by way of his interest in the anatomy of sharks and in curiously-shaped rocks known as tongue-stones.

Tongue-stones had roughly triangular shapes and could be up to several inches in length. Most of them came from Malta, where they were dug up from the ground and sold. In Europe, they were valued for their supposed curative powers; their power was considered an antidote to poison. Fishermen had noted that they were shaped a lot like the teeth of sharks.

Steno studied Maltese tongue-stones and carefully compared them to the teeth of sharks. He noticed that specific signs of wear that were common on shark teeth also appeared on tongue-stones. He concluded that the resemblance was not just approximate; the tongue-stones were petrified teeth from actual sharks. These were not "sports of nature."

Moreover, Steno saw that a similar type of petrification could help explain the many shell and fish shapes found in layers of rock. He noted that, according to the Bible, all things had been covered by water, not just once, but twice: first at the beginning of Creation, then at Noah's Flood. He speculated that when remains of animals fall into sediment underwater, the sediment may gradually harden. He noted that sudden changes in the height of land had taken place during earthquakes; these could easily re-orient and shift layers.

In 1668, Steno published his *Prodromus* [Introduction] *to a Dissertation on Solids Naturally Enclosed in Solids*. In this work he asked: Given a solid body enclosed by another, which became hard first? The first one to harden is the one that left the impression on the other. He also argued that objects that have their origin in living things may become petrified or have their shapes recorded in rock. The *Prodromus* presented Steno's three principles of strata:

1. Layers of strata have accumulated layer by layer and not all at once. First the bottom layer is formed, then the one above it, etc. This is called the "principle of superposition of strata."
2. The layers originally were horizontal or close to it.
3. A stratum forms a continuous solid layer when it is created.

Steno made no strong statements about how much time was required for the solidification of fossils; he was not contradicting the conventional biblical timeline. The *Prodromus* was approved for publication by the Church censors and published in Florence in 1668.

Steno had converted to Catholicism in 1667, and after the *Prodromus* was published, his interests turned away from science and toward religion: He remarked that "Beautiful is that which we see, more beautiful that which we know, but by far the most beautiful that which we do not comprehend."[57] Steno was ordained a Catholic priest in 1675, and his scientific work came to an end.

An English translation of Steno's work came out soon after its publication in Florence, and it was immediately influential on British scientists. One of those in England who must have read Steno's work with interest was the multi-talented Robert Hooke (1635–1703), who had addressed some of the same issues in his Royal Society lectures and had arrived at similar conclusions. The title of Hooke's geology presentation was: *Lectures and discourses of earthquakes and subterraneous eruptions, explicating the causes of the rugged and uneven face of the Earth, and what reasons may be given for the frequent finding of shells and other sea and land petrified substances, scattered over the whole terrestrial superficies.*

Hooke pointed out evidence that water could create stony substances when operating very slowly over time. Stalactites form where water drops from the roofs of limestone caverns, and living corals build stony skeletons from seawater. Similarly, Hooke argued, submerged things had become petrified to become the fossils we see today.

By 1700, due primarily to the influence of Steno and Hooke, most natural philosophers accepted that fossils came from living organisms. Although both Steno and Hooke accepted the strict biblical chronology, both had taken initial steps that would eventually lead to its downfall.

Thomas Burnet's *Sacred Theory of the Earth*

As THE NEW science of the earth started developing and becoming well known, numerous thinkers worked to integrate their new scientific knowledge with the Bible. One of the first significant attempts to develop a unifying theory bringing together the Bible with recent earth-science findings came from Anglican priest Thomas Burnet (1635–1715). In response to the Baconian injunction to keep the Bible separate from science, Burnet asserted, with unassailable logic, that no "truth concerning the Natural World can be an Enemy to Religion; for Truth cannot be an Enemy to Truth, God is not divided against himself."[58]

Burnet's central concern was to relate the history of the earth with the Biblical narrative, especially Creation, the Flood, and the Last Judgment. The biblical Flood, he estimated, took place about 1,600 years after the creation of the earth around 4,000 B.C. From the Bible, he knew that the earth as originally created was perfect, and that it would come to a fiery end at some point in the future. He also accepted the widespread view that the earth was continually degenerating.

Many held that since the Flood was a miraculous event directly caused by God, then there was no point in trying to explain where the waters had come from or where they went after the Flood ended. But Burnet had been impressed by the Scientific Revolution and its deep respect for the law of causality. He didn't like the idea that God simply created the water by a miracle and then annihilated it when He was done. Burnet wanted to explain the Flood in scientific terms.

He reasoned that even forty days and nights of the heaviest torrent could not produce enough water for the Flood, which covered the earth to a depth of 15 cubits (about 23 feet), so he asked, from where did the water come? Burnet proposed that the earth was initially perfectly spherical, without significant mountains and valleys, and that there were large reservoirs of water beneath the top crust of the earth. When the Flood occurred, parts of the crust split, and rocks of the crust collapsed into the reservoirs, forcing the water up to the surface. When the Flood ended, the water seeped back into gaps in the earth but left the surface in the chaotic state we see today.

Burnet published his *Sacred Theory of the Earth* in 1681 in Latin, and shortly after that in an English version. The full title was *The Sacred Theory of the Earth: Containing an Account of the Original of the Earth, and of the General Changes which it has Already Undergone, or it is to Undergo, Till the Consummation of All Things*.

In 1692, Burnet published his *Archaeologiae Philosophica* (Philosophic Archeology), which tried to reconcile his theory with Genesis in greater detail, treating the biblical account of the fall of man as an allegory, not as literal truth. This led to accusations of blasphemy, and Burnet was forced to resign his position.

Burnet's *Sacred Theory of the Earth* inspired many similar works, such as the 1696 book by professor William Whiston (who would later succeed Isaac Newton as Lucasian Professor of Mathematics at Cambridge) titled *A new theory of the Earth, from its Original to the Consummation of All Things, wherein the Creation of the World in six days, the universal deluge, and the general conflagration, as laid down in the Holy Scriptures, are shewn to be perfectly agreeable to reason and philosophy*.

In the early eighteenth century, a handful of thinkers on the Continent, influenced by the philosopher René Descartes, were starting to consider much longer time frames for the age of the earth. Descartes had proposed a mechanistic view of the physical universe, in which the earth had formed from the material of an extinguished star. Descartes did not discuss the duration of this process, but he did not restrict himself to an account literally consistent with Genesis.

A French diplomat and traveler named Benoit de Maillet (1656–1738) was impressed by evidence suggesting that sea levels had been steadily decreasing over time. Some ancient inland cities must have surely been seaports in their day, but these were now far above sea level. At a more recent seaside fortress at Carthage, markings indicating the initial water level are now 5–6 feet above sea level. De Maillet reasoned that the sea must be sinking about three inches per century. He argued that the diminution of the sea must have started at least two billion years ago.

At some point in the distant past, according to De Maillet, the earth was covered by water. This water has been gradually evaporating and disappearing. Echoing Anaximander, he argued that all life arose in the sea, and had gradually transformed into the living world we see today.

De Maillet did not restrict himself to the Biblical timescale, or even attempt to reconcile his views with Genesis. He revised his book continually until shortly before his death in 1738. It was not published until 1748, as *Telliamed: or Conversations between an Indian philosopher and a French missionary on the diminution of the sea*. The editor made many changes to make it less heretical, but it was still violently denounced. However, the book sold well and an English translation was made in 1750.

Another Frenchman who was open to a longer-than-biblical timescale was Georges Leclerc, Comte de Buffon (1707–1788), Keeper of the Jardin du Roi for Louis XV, and author of the multi-volume *Histoire Naturelle, générale et particulière* (1749–1804). Most of this encyclopedic work was devoted to descriptions of existing plants and animals, but Buffon also presented his theories about geology. He discussed the processes of mountain-building, sedimentation of river estuaries, erosion of coasts, the cooling of the earth, the formation of rocks and minerals, ocean and atmosphere, and the emergence of life—all as natural processes, occurring over at least 100,000 years. Members of the Sorbonne's Faculty of Theology (representatives of the Catholic Church) denounced Buffon's work and sent him an official letter identifying a list of propositions (both philosophic and scientific) from the work that "seemed reprehensible" to them. The first four propositions involved his speculations on how the earth might have been formed from the sun. Buffon was forced to publish a statement that he never had "any intention of contradicting the text of Scripture" and that he abandoned anything in his book that "touches on the formation of the Earth, and in general everything that might be contrary to the narrative of Moses."[59]

The Cartesians Benoit de Maillet and the Comte de Buffon had opened the door to the possibility of a genuinely ancient earth and the corresponding rejection of the literal Genesis account. Both had made intriguing observations, but their writings were still speculative, unconvincing, and easily dismissed by critics. The next major development was to come from the Scottish Enlightenment, in the person of a gentleman farmer named James Hutton, who would become acclaimed as the father of modern geology.

James Hutton's Deist Geology

JAMES HUTTON WAS born in Edinburgh in 1726, son to a merchant and city treasurer. Hutton studied at the University of Edinburgh, where he developed an interest in chemistry. He also attended lectures on mathematics by Colin Maclaurin—an accomplished mathematician who had met and been deeply influenced by Isaac Newton, and who would go on to extend Newton's work in calculus, geometry, and gravitation.

Maclaurin taught that this world is not the work of the Devil or a decayed ruin of a more perfect past time, as was conventionally believed. Rather, he held that it is an orderly and benevolent place that reflects the generosity of God and the symmetry of his natural laws. His deism was a significant step away from traditional Christianity, in that deists accepted the moral guidance of the Bible, but rejected its literal interpretation as an accurate history of the world. God set the universal laws, but did not tinker with these laws once they were set; He definitely did not intervene in the world through miracles and floods. The deist philosophy would profoundly influence Hutton.

Hutton received his medical degree from the University of Leiden. Having developed an avid interest in chemistry while back in Edinburgh, he started a business with a friend for manufacturing sal ammoniac from chimney soot. (Sal ammoniac is a white salt that was used in dyeing and for working with brass and tin.) After inheriting a farm in Scotland, he decided to become a farmer; he decided to learn all he could about different farming techniques, and this led him to explore much of England and the low countries of Europe. During this exploring he became fascinated by the structure of the earth's surface, and he started developing his own theory around 1760. In 1768 he retired from farming and moved to Edinburgh, where he further developed and tested his ideas.

Hutton became so familiar with the geological structures and rocks of England that he would boast that he could tell where a piece of gravel had come from anywhere on the eastern side of Britain. A friend of Hutton owned and operated a small coal mine, and this gave Hutton more opportunities to study the structure of the rocks. Over his lifetime Hutton amassed a huge collection of geological specimens, many collected by himself, and many more acquired from a network of like-minded associates. He also devoured travel books for any information about geology.

Hutton noted that the most obvious type of geological change was erosion by water. All farmers knew the power of water to wash away soil, but Hutton could see that erosion affected rocks as well. At places on the coast, he saw the effects of waves on rock outcrops that faced the sea. He concluded that most valleys had been created by the erosion of streams. Hutton saw a pattern of destruction of the land, with rocks slowly decaying and disintegrating into pieces that comprise soil, which is later washed away by streams and rivers, eventually leading to the sea. If this process continued indefinitely, he reasoned, the continents would eventually be leveled and swept entirely into the sea.

Belief in the "degeneration" of the land was relatively common in Hutton's time. Many earlier thinkers such as Thomas Burnet believed that the earth itself is in a state of decay, having been created initially in a perfect state, but having been affected by Adam's Fall and Noah's Flood. Now its surface is in a disordered and messy state.

The idea of a decaying earth was not entirely consistent with Hutton's deist philosophy. God had created a finely tuned world-machine for the sustenance of life, and this machine could not wear out, run down, and decay. Or if it did have elements that necessarily decay, it would also have inbuilt means of revitalization. The world must surely sustain itself. But how?

Hutton saw indications that rocks had been created. He examined the small-scale structure of sedimentary rocks such as sandstones, and compared this to the gravel and sand produced by erosion; they looked the same, except that in the rocks the particles had somehow been cemented together. As a farmer, Hutton was well aware of how farmers throughout southern England used chalk as a fertilizer. On close examination, chalk seemed to contain the bodies of sea animals.

Hutton could see that erosion destroyed rock, unknown forces created rock, and within this process was created the soil necessary to support life. This was not explained by any existing geological theory.

Hutton accepted the conclusions of Hooke and Steno about how rock layers had formed. But at the bottom of the sea, how could layers of sediment become hardened? In Edinburgh, Hutton became acquainted with chemist Joseph Black, whose experiments with quicklime demonstrated that heat and pressure could transform rocks in various ways. Hutton concluded that the layered rocks had been hardened by the combination of pressure from above layers, combined with heat from the earth.

Since the layers were often found far from the seashore, either the sea had dropped or the land had been pushed up from below. In many places the strata we see are nearly horizontal, but this is not the case through much of Scotland, where the strata can be found at almost any angle, clearly bent, and even turned over. Some tremendous pressures in the earth must have bent the strata, and if they could *bend* the strata, they could also *raise* the strata.

Hutton was aware of signs of heat deep in the earth, such as volcanoes and hot springs. He had heard about earthquakes which had caused land masses to rise. At places on the coast he saw indications that granite and basalt had been injected, in a molten form, in veins into previously existing strata.[160] The inventor James Watt, a friend of Hutton, had shown that heat could do work; so, heat could conceivably cause land to rise.

Hutton concluded that the surface of the earth had changed cyclically. Land was worn away into particles that were deposited into the sea in layers, which were then baked by the earth's heat and then pushed up into new land, and the cycle continued indefinitely.

From this theory, Hutton predicted that there would be places where two different sets of strata, oriented in different directions, are in contact. After an extensive search, Hutton was able to find actual instances of these structures, which are called "unconformities" by modern geologists.[161] These showed that the first set of strata were laid down in a sea, hardened, then lifted and deformed, and eroded, then submerged, and new strata were laid on top of that. This confirmed Hutton's idea of cycles of uplift and decline.

So how old is the earth? How fast are the processes of creation and decay of rocks, and how many times has this happened in the past? Hutton did not have answers to these questions. He did hold that the earth must be extremely ancient. Consider how slow the erosion of rocks must be. Hutton knew that Hadrian's wall across Britain was about 1600 years old, yet parts of it were still in good condition. The process of decline was too slow to be measurable within the timescale of human history:

> Every revolution of the globe wears away some part of some rock upon some coast; but the quantity of that decrease, in that measured time, is not a measurable thing. . . .
>
> We are certain, that all the coasts of the present continents are wasted by the sea, and constantly wearing away upon the whole;

but this operation is so extremely slow, that we cannot find a measure of the quantity in order to form an estimate.[162]

Hutton concluded his 1788 paper saying that, for the world, we can find "no vestige of a beginning, no trace of an end." He did not mean that the earth is eternal, although many interpreted that to be his position.

Hutton first publicly presented his views in 1785, in an address to the Royal Society of Edinburgh. He later summarized his theory in a concise essay, published in 1788, in the *Transactions of the Royal Society of Edinburgh*. His work culminated in 1795 with his publication of the two-volume *Theory of the Earth with Proofs and Illustrations*, which was over 1,100 pages in total and written in a dense style that did not make it easy reading.

The reaction to Hutton's theory was mixed. Some journal reviews were moderately positive, but many geologists simply ignored his theory, and others attacked it mercilessly. In their attacks, the dominant issue, though not always named explicitly, was the conflict with Genesis.

Hutton's theory clashed with the dominant geological view known as Neptunism. Named after the Greek god of the sea, this theory had been propounded most strongly by lecturer Abraham Werner. Neptunists held that heat was an insignificant agent for the creation of rocks, and that virtually all rocks had been precipitated or crystallized out of a universal ocean. They asked Hutton's followers how the earth could contain a virtually-limitless source of heat (which seemed necessary for Hutton's theory). Neptunism could be much more easily reconciled with the Bible. Werner's ideas became prevalent in Europe in the late eighteenth century and into the nineteenth. While Werner himself did not hold to the biblical chronology, many other Neptunists did.

Richard Kirwan of Edinburgh, a respected chemist and geologist, explicitly attacked Hutton's theory on two grounds: it created an incomprehensible abyss of time, and it was contrary to Genesis. John Williams, the author of a 1789 book titled *The Natural History of the Mineral Kingdom*, devoted forty pages to an attack on Hutton's theory, calling it a "wild and unnatural notion" that led to "scepticism, and at last to downright infidelity and atheism."[163] Numerous churchmen also chimed in on the debate, and their opinions were as expected: In 1813 Reverend Joseph Townsend published his *The Character of Moses Established for Veracity as an Historian, Recording Events from the Creation to the Deluge.*

Although Hutton's deism and appeal to divine-design arguments played a role in his reasoning, his primary sources were the many types of evidence he presented. His reasoning was, overall, dominantly empirical. His *Theory of the Earth* is one of the most influential books ever published in geology; his fame as the "father of geology" is deserved. He was the first to create a systematic and testable theory of the earth that rejected any Bible-based reasoning. He had discovered *deep time*.

Charles Lyell

AFTER HUTTON, THE next giant of geology was another Scotsman. Charles Lyell (1797–1875) studied at Exeter College, Oxford, where classics and theology dominated the curriculum. Lyell attended the lectures on geology of the Reverend William Buckland, who contended that geology pointed to a divine Deluge. Geology was becoming a popular and exciting subject, in which many were working to integrate the new scientific discoveries with Genesis. When Lyell read a book on Hutton's theory, he became convinced of its essential truth. He spent the rest of his life studying and writing about geology, traveling extensively through Europe, making countless observations, and reading works relevant to geology, in numerous languages.

Lyell was profoundly influenced by Hutton's theory, but unlike Hutton, Lyell did not rely on arguments from design; he emphasized three closely-related issues of method:

1. As demonstrated by Newton's *Principia,* unchanging universal laws of nature are the key to understanding the world.
2. In geology, "the present is the key to the past."[164] We should always attempt to explain past changes by reference to the processes that we see operating today.

3. We should generally assume that the intensity of geological processes of the past (including earthquakes, volcanoes, floods, etc.)
is similar to the intensity we observe in the present. Positing
unobserved past catastrophes (such as a universal deluge) is too
speculative to be scientific.

Lyell's magnum opus was titled *Principles of Geology: Being an Attempt
to Explain the Former Changes of the Earth's Surface, by Reference to Causes
now in Operation* was published in London, in three volumes, between
1830 and 1833. The 1,400 pages of the text were rich with examples of geological forms and changes over time. His research made use of hundreds
of sources in five modern European languages and two ancient ones.

Most of the text was devoted to describing observed geological
forces. Lyell carefully described the slow changes involved with rivers, tides, currents, and delta-formation. He wrote about the effects of
earthquakes and volcanoes, as well as geological changes resulting from
plants and animals. He devoted long chapters to the description of fossils and the processes that must have created them.

Lyell saw the Anglican Church as the main obstacle to a real sci-
ence of geology; most Anglicans were committed to a literal reading of
Genesis. Lyell passionately desired to "free the science from Moses."[165]
But he did not attack Genesis outright. Lyell reassured his conserva-
tive readers that geology was not anti-Bible or anti-Christian and that
Genesis was open to a less-than-literal reading. The first six days of
Genesis were not necessarily six literal days, as even Saint Augustine
had accepted; they may have instead been six eras of unknown length.
Geology, Lyell also assured his readers, did not challenge the biblical
story of the origin of man.

Lyell's *Principles of Geology* was widely read, discussed, and reviewed,
in Europe, America, and Australia. Lyell constantly revised his *Principles
of Geology* to account for recent discoveries. When he died in 1875 he was
busy revising the twelfth edition.

In early nineteenth-century Britain, most people still weighed the
literal truth of scripture higher than science-based accounts. Books like

Burnet's were popular through the first half of the century. But Lyell's books were profoundly influential, and by the mid-nineteenth century, most educated Englishmen had opened to the possibility that the earth was indescribably old. The new geological conclusions still alarmed some (mainly clerics), but in general, these discoveries were not considered a major threat to religion.

The most significant impact of Lyell's writings is evident in the writings of Charles Darwin, who took the first volume of the *Principles of Geology* along on his H.M.S. Beagle voyage. Darwin wrote to his sister that before reading Lyell he was ignorant of geology: "As far as I know everyone has yet thought that the six thousand odd years [of Ussher] has been the right period."[166] Darwin's debt to Lyell (and the earlier geologists such as Hutton) is evident in Darwin's masterwork *On the Origin of Species*:

> It may be objected, that time will not have sufficed for so great an amount of organic change, all changes having been effected very slowly through natural selection. . . . He who can read Sir Charles Lyell's grand work on the Principles of Geology, which the future historian will recognise as having produced a revolution in natural science, yet does not admit how incomprehensively vast have been the past periods of time, may at once close this volume.[167]

* * *

In this chapter, we have seen how scientific observation and reasoning led thinkers to reject the Biblical view of Earth's history and to embrace the idea of *deep time*. The discovery of deep time eroded belief in the literal truth of the Bible and helped pave the way for Darwin's theory of evolution. Although earlier thinkers had argued for deep time, the single most important figure was James Hutton.

This was a transition away from reliance on the Bible (a product of consciousness) as a source of knowledge. As such, it was a step away from the *primacy of consciousness* and towards the *primacy of existence*.

CHAPTER ELEVEN
Darwin and Evolution: Genesis Under Attack

I N THE LAST chapter, we saw how discoveries in geology undermined the literal account of Genesis. By the mid-nineteenth century, it was widely accepted that the story of the creation of the world in six days was not literally true, that the world developed significantly over vast periods of time. This made it conceivable that living things—and perhaps man himself—might also have changed.

William Paley's *Natural Theology*

IN CHAPTER SIX, we saw how the study of natural theology flourished in England in the seventeenth century. The interest continued into the nineteenth century. It acquired a strong new defender in the person of the Reverend William Paley, who published an enormously popular book in 1802 titled *Natural Theology: or, Evidences of the Existence and Attributes of the Deity, Collected from the Appearances of Nature.* The book opens as follows:

> In crossing a heath, suppose I pitched my foot against a stone, and were I asked how the stone came to be there, I might possibly answer that, for any thing I knew to the contrary, it had lain there for ever; nor would it perhaps be very easy to shew the absurdity of this error. But suppose I had found a watch upon the ground, and it should be inquired how the watch happened to be in that place. . . . [W]hen we come to inspect the watch, we perceive, (what we could not discover in the stone,) that its several parts

> are framed and put together for a purpose, e.g. that they are so
> formed and adjusted as to produce motion, and that motion so
> regulated as to point out the hour of the day; that, if the several
> parts had been differently shaped from what they are, of a differ-
> ent size from what they are, or placed after any other manner, or
> in any other order . . . none would have answered the use, that
> is now served by it. . . . [T]he inference, we think, is inevitable;
> that the watch must have had a maker; that there must have
> existed, at some time and at some place or other, an artificer or
> artificers . . .[168]

Paley makes an argument from analogy in which the watch is anal-
ogous to the living organisms we see around us. This is a form of the
"argument from design" for the existence of God. The argument from
design had been presented by numerous earlier thinkers (such as
Thomas Aquinas), but Paley gave it a powerfully eloquent statement,
designed to appeal to the age of reason, with its respect for our powers
of observation and logic. *Natural Theology* contains no mentions of miracles or references to the Bible.

Although many thinkers (including the young Charles Darwin)
found this argument convincing, the "argument from design" is not logically valid. Ignorance is not evidence for the supernatural. Ignorance of
the natural cause of life is not evidence that there is no natural cause.
Ignorance about the natural world is not evidence for some realm
beyond the natural world. Nevertheless, this argument was convincing
for many nineteenth-century thinkers. The idea that, in principle, "a
natural explanation may someday be found for the organization of living
things" was not satisfying to them.

Pre-Darwin Views of Evolution

Before the nineteenth century, there were only a handful of thinkers who had considered the possibility that current living things had
evolved from earlier forms. In Ancient Greece, Anaximander (c. 610–c.

546 B.C.) speculated that all animals (including man) initially emerged out of the sea. During the eighteenth century, Enlightenment materialists such as Denis Diderot and Baron d'Holbach proposed that non-living matter, when in the right conditions and combinations, could give rise to living things, even complex ones. They were fascinated by the existence of "monstrous" births, such as a two-headed calf. Whereas in the Middle Ages such births had been considered signs from God, Diderot and d'Holbach considered them rare but natural events, which could lead to new species.

Another Enlightenment figure who speculated about evolution was the deist Erasmus Darwin, a successful doctor and the grandfather of Charles Darwin. Erasmus Darwin proposed that God had designed living things to be somehow self-improving and to adapt to their environment. He presented his ideas in a chapter of his medical book *Zoonomia: or the Laws of Organic Life* (1794).

The first theory of evolution to rise above the level of sheer speculation was that of Jean-Baptiste Pierre Antoine de Monet, chevalier de Lamarck (1744–1829). As chair of invertebrate studies at the French Museum of Natural History, Lamarck was impressed by the fact that living organisms exist on a continuum of levels of complexity, as well as being adapted to their environments. The geologic column seemed to indicate a sequence from simple to more complex.

Lamarck first published his theory in an 1802 paper and then elaborated it in his 1809 treatise *Zoological Philosophy*. According to his theory, the simplest living organisms are spontaneously generated out of physical matter when it is acted upon by a material life force or fluid. As historian Eric Larson summarizes:

> This force or fluid could transform "gelatinous" matter into the simplest of animals, he claimed, and "gummy" matter into the simplest of plants. Once living organisms form (and this happens continuously, according to Lamarck), the fluid continues to act on them and their descendants—naturally driving them to evolve into ever more specialized forms.[169]

Within an organism, Lamarck held that the vital fluid naturally flows toward used organs and away from unused ones, causing the former to develop further and the latter to atrophy. Also, he held, like Erasmus Darwin, that acquired characteristics could be inherited. As a giraffe repeatedly stretches its neck to reach leaves on a tree, the life fluid flows into its neck muscles and causes it to get a little longer. Then this longer neck would somehow be inherited by the giraffe's offspring.

Lamarck's theory of evolution encountered immediate and strong opposition from other naturalists. The most strident critic was the accomplished scientist Georges Cuvier (1769–1832), often considered the founder of modern comparative anatomy and paleontology. The French Royal collection of animals and fossils was probably the most extensive such collection in the world, and Cuvier had made good use of it, to become the acknowledged master of comparative anatomy.

As a result of his studies, Cuvier concluded that none of the species preserved in fossils existed anymore, so they must have died out. He was the first scientist to conclusively establish that many animals had gone extinct. (Before Cuvier, most people believed that no species ever died out, and that fossils were not meaningful.) Cuvier also concluded that every species is ideally suited for its environment, and unchanging, with each of its parts perfectly fitted to the others.

Cuvier objected to Lamarck's theory on the grounds that it was wildly speculative, with no real evidence behind it. It was not consistent with the many gaps in the fossil record, and there was no evidence for Lamarck's "vital fluid" or the inheritance of acquired characteristics. But Cuvier's main argument relied on his firm belief that every part of every organism is perfectly and precisely integrated into the organism, so that any small change would make the organism non-viable. This view was not consistent with evolution (unless all of an organism's parts evolved simultaneously).

If Cuvier did not believe in evolution, how did he explain the fossil record? Cuvier concluded that there had been a series of catastrophes on the earth's surface, which had led to mass extinctions. After each of these, there had been repopulations of organisms perfectly suited to these new environments. These were possibly migrations from elsewhere, but some of these must have been new creations. Implicit in Cuvier's theory was a dependence on God for creating new species, but as a student of the Enlightenment, Cuvier never gave scientific arguments based on the Bible.

Whereas Lamarck and Cuvier wrote primarily for their fellow naturalists, the author of the next proposed theory of evolution wrote for the general reading public. Scottish writer Robert Chambers anonymously published in 1844 his *Vestiges of the Natural History of Creation*, and it sold well among the British public.

Chambers argued that evolution occurred as part of a divine plan of progress. It began with spontaneous generation on an initially lifeless earth. Then there was a natural progression from simple living forms all the way to humans, with side branches of the tree leading to other animals and plants. The overall pattern of development was preordained by God. No miracles were required because God had incorporated the necessary laws of creation into the structure of the universe. Unlike Lamarck, Chambers did not specify any mechanism for the adaptation of an organism to its environment; in fact, he barely spoke of adaptation at all.

Naturalists heaped scorn on Chambers' book, as he had no real evidence to support his claims. But the book was widely read by the general public, and many started warming to the possibility of evolution.

Darwin's Discovery of Natural Selection

As a youth and young man, Charles Darwin (1809–1882) was utterly fascinated with the natural world around him, and he collected and cataloged biological specimens with a dedication unusual for a boy his age.

Despite his undeniable passion for biology, Darwin did not immediately choose it as his calling. Following in his father's footsteps, he first studied medicine. Then he turned to theology, but revealingly, the details of the Anglican faith didn't hold his interest. During all of his studies, however, he had always made time for his true love—natural history. He took university courses on the subject, and his dedication gained him the admiration of a professor of natural history at Cambridge.

This professor was instrumental in getting Darwin a position on the survey ship H.M.S. Beagle. The young captain was looking for an educated travel companion of his own age, and Darwin fit the bill perfectly. The ship was to survey numerous locales throughout the southern hemisphere, and the journey was to last several years. When Darwin returned to England almost five years later, he had amassed a staggering collection of observations—and enough specimens to fill a museum:

> He had left England as a young man fresh out of university with only a good upbringing, a deep-rooted enthusiasm, and amateur knowledge of geology and naturalism to his credit. He was returning with 1,383 pages of geology notes, 368 pages of zoology notes, a catalogue of 1,529 species in spirits and 3,907 labeled skins, bones, and miscellaneous specimens, as well as a live baby tortoise from the Galapagos Islands. His diary amounted to 770 pages, and parts of his journals, sent ahead of him, had already been read by a number of scientists at home.[170]

Due to Darwin's lifelong dedication as a naturalist, he became intimately familiar with a wide range of natural phenomena, from the most minute concretes to the widest possible abstractions—from such details as the subtle differences in the beaks of various Galapagos finches to the hierarchy of similarities and differences relating all living organisms. His thinking on the question of evolution was grounded not in armchair speculation, but his vast knowledge of all of life's varied manifestations.

Darwin was also asking probing and unprecedented questions. If God had designed each creature as a perfect fit for its environment (according to the conventional view), why do identical environments that are geographically separated (such as the Galapagos Islands) support different species? And why do the species of South America (which contains vastly different environments) share essential similarities with each other, which they don't share with their counterparts in Africa and Asia?

But the most crucial question Darwin struggled with was that of how evolution occurs. Unsatisfied by his predecessors' vague notions about a divine plan or a natural striving, he sought a causal mechanism for evolution that had a discernible basis in observable facts. His answer—the theory of natural selection—was inspired, ironically, by a profoundly flawed treatise on human population growth: Thomas Malthus's *Essay on Population* (first published in 1798). Malthus maintained that human populations tend to increase "geometrically" (exponentially), while food and other natural resources only increase "arithmetically" (linearly). So, argued Malthus, with population growth outpacing food production there is a natural tendency for us to reproduce to the point of starvation.[171] Although this argument fails when applied to human beings (whose rational faculty trumps natural population pressures), it stimulated Darwin to consider its relevance for animals:

> The message that Malthus' *Essay* brought home to Darwin is that the majority of all individuals, in the natural state, do not survive long enough to reproduce. Darwin wondered why some individuals, the minority, should survive and reproduce, while others did not. And he saw that the survivors would be the ones best suited to the way of life of that species—best fitted to their ecological niches, in the way that a key fits into a lock, or a piece fits into its place in a jigsaw puzzle.[172]

Darwin's revolutionary idea was that this process of natural selection would act on naturally occurring variations among individuals of the same species. This would provide a genuine causal mechanism for evolution, and it could, in principle, be tested. Darwin wrote later in his *Autobiography*: "I now had a theory with which to work."[173] He resolved to gather more evidence and see if it fit the theory. He decided to study the barnacle—a species whose classification was poorly understood at the time. This developed into a massive undertaking, but it reinforced his confidence in natural selection, because key to this theory was the existence of a large amount of variation among the members of a species.

From Darwin's intensive study of barnacles, he concluded that variety was not the exception, but the rule. There was no exact body plan governing all members of the same species; every aspect of every species could be found modified to some extent. As Darwin wrote to a friend, "I have been struck by the variability of every part in some slight degree of every species when the same organ is rigorously compared in many individuals."[174] Even a feature as seemingly clear-cut as sex was variable. Barnacles were generally hermaphroditic, with the same organism possessing both male and female sex organs. But there were also species in which male and female were separate organisms that lived attached to each other, or in which several males were attached to a single female.

Darwin spent many years patiently working out the details of his theory and gathering more and more evidence to bolster it. He realized that his new ideas constituted an outright assault on the accepted religious view of the origin of species, so he was apprehensive about going public. He shared his ideas on evolution with only his closest friends while he continued to gather evidence.

Darwin was, in effect, forced to publish when a young naturalist, Alfred Russell Wallace, discovered the same basic idea independently. Wallace had written a letter to Darwin asking his opinion of the theory. The letter was a complete shock to Darwin, and he was ready to grant Wallace full priority to the discovery. But Darwin's friends helped him coordinate a joint publication, so the theories of both men were first published together, as short papers.

Darwin's magnum opus, *On the Origin of Species by Means of Natural Selection*, was published in 1859. It was an instant best-seller. Written for a general audience, it laid out a clear argument, starting with "artificial selection." Many people knew that animals and crops had been significantly modified by careful human intervention, by selection of the individuals with the desired traits for breeding. Darwin then proceeded to the idea of "natural selection," along with the evidence that it had occurred.

Darwin presented the fossil sequence of the geologic column. He also proceeded to give the evidence from biogeography (the geographic distribution of species), anatomy, and embryology. The early embryos of a human and a dog look almost the same. Darwin's theory of natural selection made sense of rudimentary organs (such as the human tailbone) and homologous correspondences in comparative anatomy (such as the five-fingered bone structure of mammalian hands, paddles, and wings). These made perfect sense as by-products of natural evolution, but not of a mind designing each species to perfectly fit its environment. Although the book did not explicitly say that *humans* had descended from lower forms, the implication was clear and was made explicit in Darwin's later work, *Descent of Man*. The question of our origins is particularly revealing of Darwin's extraordinary intellectual honesty. Unlike Wallace, whose belief in our divine creation prevented him from ever accepting our evolutionary origins, Darwin refused to place any consideration higher than the truth. He was willing to follow the facts wherever they led—even if that required him to boldly confront the religious dogmas of his day. And there is no better sign of Darwin's total commitment to reason than the fact that the former student of theology died a nonreligious agnostic.[175]

The theory of evolution by natural selection is a historic example of scientific induction on a grand scale—a vast integration concerning the nature and development of every living organism. What made it possible was Darwin's lifelong dedication to understanding the natural world, from the smallest to the grandest scales—and his total devotion to truth.

The Temporary Waning of Natural Selection

DARWIN'S *Origin of Species* convinced many of the reality of evolution; however, it was not as effective in converting others to *natural selection* as the mechanism of evolution. In fact, after an initial swelling of support, there was a decline of interest in natural selection as various critiques were published. The most effective of these critiques came from the science of physics.

In the 1860s the great physicist William Thomson (later Lord Kelvin) calculated the age of the earth, using the science of thermodynamics, based on the assumption that the earth was initially a ball of entirely molten rock that had been gradually dissipating its heat into space. His initial estimate was under one hundred million years, and as he refined his calculation, his estimates went down. (Thomson's calculations did not take into account the radioactivity of the earth, of which he was not aware.) These estimates were generally acknowledged, even by Darwin, as not long enough for natural selection (on its own) to have created all current living things.

Given the fact that Lamarckian evolution, if it occurred, would operate much faster than natural selection, numerous biologists started reconsidering Lamarckian-like theories as alternative or supplemental to natural selection. These included even Darwin himself, who proposed a theory of "pangenesis" which hypothesized a mechanical explanation for the inheritance of acquired characteristics. According to this theory, every part of an organism generates tiny "gemmules" containing hereditary information about itself. Some of these gemmules find their way into the ova and sperm, through which they may be passed to offspring.

Other mechanisms for evolution were also proposed. "Orthogenesis" was the theory that once an organ started evolving in a particular direction due to natural selection, then there was a type of "momentum" which caused it to continue evolving even when there was no survival benefit to the continuing changes.

The theory of evolution by saltation (also called mutation theory) held that, occasionally, an organism would give birth to a mutated offspring that would start an entirely new species. While saltation theorists acknowledged that natural selection would kill off clearly harmful mutations, they often did not see natural selection as playing a primary role in directing evolution.

There was also a theory of "theistic evolution" as seen in the writings of American botanist Asa Gray (1810–1888). Gray agreed with the survival-of-the-fittest aspect of Darwin's theory, but held that God directly causes the variations on which that selection acts.

In the 1860s, the Augustinian monk Gregor Mendel had published the results of his hybridization experiments with pea plants, leading to the laws of heredity for binary hereditable traits (which could take one of two forms, such as yellow vs. green, or round vs. wrinkled). Unfortunately, the significance of Mendel's work was not appreciated at this time, since it was not clear how they related to continuous traits

Darwin and Evolution: Genesis Under Attack

such as the length of a giraffe's neck. At this time, most biologists had a "blending" view of inheritance, in which they saw an offspring's characteristics as a "blended" version of its parents' characteristics.

Mendel's work was forgotten for decades, until around the year 1900, when it was rediscovered by three different biologists, who then restarted the science of genetics. Ironically, these biologists did not see their work as compatible with Darwin's theory of evolution. Genes seemed to control discrete and "big" differences, not continuous traits.

The first large-scale experimental study of genetics was started by Thomas Hunt Morgan at Columbia University in New York in the 1910s. The organism he decided to use was the fruit fly, *Drosophila melanogaster*. It was inexpensive to feed and maintain, easy to breed in large numbers, and created a new generation every twelve days. Morgan focused on the process by which the accumulation of random, inborn mutations over time can lead to new types of organisms. He saw natural selection as only a process of weeding out harmful mutations, not as leading to adapting an organism for its environment. Like Mendel, Morgan studied discrete traits, in which substantial changes seemed to occur suddenly.

The Neo-Darwinian Synthesis

THE SYNTHESIS OF genetics with natural selection began in the 1920s with the work of British biochemist J.B.S. Haldane. In a series of highly mathematical papers published in the later 1920s and early 1930s, he argued that Darwin's natural selection combined with Mendelian genetics yields a full mechanism for evolution. Haldane presented mathematical arguments showing that varieties possessing even a slight competitive advantage would eventually come to predominate within a population, due to a process similar to compound interest in banking.

In 1937, the Russian émigré Theodosius Dobzhansky published his book, *Genetics and the Origin of Species*, which for the first time presented the new synthesis for a general audience. By mid-century, the neo-Darwinian synthesis had become almost universally accepted among biologists. The theory generated a wide variety of predictions that were later verified (such as the existence of specific intermediate fossil forms). This overwhelming explanatory success led to Dobzhansky's conclusion that "Nothing in biology makes sense except in the light of evolution."[176]

The Rise of Evangelical Fundamentalism: Religion Strikes Back

DARWIN'S THEORY OF evolution prompted a range of different reactions, corresponding to the various ways in which people tried to integrate it with their views about religion. At one end of the spectrum were atheists like Thomas Henry Huxley, who saw Darwin as banishing God from the world. But this view was relatively rare.

More popular was the view of those, like naturalist Alfred Russell Wallace or geologist Charles Lyell, who accepted the theory's validity for all living organisms except for humans. As they argued, humans are *sui generis* among the living creatures of the world. We alone have abstract conceptual thought, free will, morality, art, and, crucially, an ineffable spiritual soul. For Wallace and Lyell, at some point during the evolution from primitive primates to modern humans, a supernatural agent must have stepped in to create our unique souls. Thinkers such as writer Robert Chambers held an intermediate position, in which he saw evolution as a fully continuous process, engineered or guided by God, with the end result of creating human beings.

The idea of evolution directly contradicted the literal story of the creation of man in Genesis, along with the story of Adam and Eve, and the source of original sin. One reaction among the religious was to retreat and start taking the Bible in a non-literal way, focusing on it more as a source of parables presenting moral ideals. Some Christian sects such as the Episcopalians and Congregationalists were more liberal and more open to evolution. In 1885, Congregational pastor Henry Ward Beecher published *Evolution and Religion*, which argued that evolution is the "method of God in the creation of the world."[177]

During the eighteenth and nineteenth centuries, there was a gradual increase in the number of scholars studying the Bible and other religious texts in a more secular framework, more as a piece of literature and history than as the revealed word of God. This was referred to as "higher criticism" (distinguished from "lower criticism," which focused on the exactness of the translations from Hebrew and Greek).

French writer Ernst Renan wrote the famous *Life of Jesus* (*Vie de Jesus*), which was translated into English in 1863. German theologian David Friedrich Strauss wrote *Das Leben Jesu, kritisch bearbeitet* (*Life of Jesus, Critically Examined*), published in German in 1846 (in 3 volumes). Both of these works treated Jesus as a purely historical figure, proposed that the miracles of the gospels were mythical, and as a result, generated a storm of controversy.

The most widespread reaction to both evolution and higher criticism was to recognize them as attacks on the literal Bible and the foundations of Christianity. In 1874, Princeton theologian Charles Hodge published his book *What is Darwinism?* His answer: "It is atheism [and] utterly inconsistent with the Scriptures."[178] Genesis clearly describes a world in which God created man in the perfect Garden of Eden, and in which man sinned against God and thereby was condemned by Him to a life of toil and pain. The evolutionary view, in contrast, held that humans had gradually risen from savage beginnings to their current civilized state.

In the United States between 1900 and 1920, government-controlled secondary-level education quickly became widespread. In areas that had previously been extremely isolated, large numbers of people suddenly learned about modern science for the first time. In their science classes, students studied Darwin's theory of evolution, and came home and told their parents that they were descended from apes. Their parents were not happy to hear this.

In the 1910s, a group of American evangelicals began calling themselves "fundamentalists." They committed themselves to what they saw as the fundamental tenets of Christianity: the inerrancy of Scripture, the veracity of Old and New Testament miracles, and the trustworthiness of end-time prophecies. Baptist Reverend A. C. Dixon edited a series of tracts, called *The Fundamentals*, between 1910 and 1915, which expounded these tenets.

In the 1920s, America's preeminent evangelist was Billy Sunday. His sermons professed the following ideas: Man's inherent sinfulness; redemption through personal faith in Jesus; the Bible as God's literal word. He attacked the scholarship of the evolutionists and those of higher criticism: "When the word of God says one thing and scholarship says another, scholarship can go to hell."[179] Sunday, in his sermons, helped to inspire America's first law against teaching evolution (in Tennessee), which was passed in 1925.

In 1923, George McCready Price published *The New Geology*, which attacked geology as the weakest point in evolution theory. He pointed out examples of the violations of the law of stratigraphy, and asserted his own law as a replacement: *the Law of Conformable Stratigraphical Sequence*: "Any kind of fossiliferous rock may occur conformably on any other kind of fossiliferous rock, old or young."[180] Price argued that the Biblical deluge could have created the entire geologic column, and some natural sorting process must have placed the organisms in the current layers.

The Scopes "Monkey" Trial

AFTER THE PASSAGE of the Tennessee law against teaching evolution in 1925, the American Civil Liberties Union (ACLU) publicly offered to defend any Tennessee teacher willing to challenge the new law. A young science teacher named John Scopes promptly accepted the offer. The trial quickly became a major publicity event, and both sides obtained famous lawyers. The prosecution was led by William Jennings Bryan, a populist politician from Nebraska (Senator and three-time presidential candidate), who was known as the "Great Commoner." Leading the defense team was the legendary criminal defense lawyer Clarence Darrow, who had in recent years spoken out against religious restrictions on personal freedom.

The prosecution called students and school officials to the stand to testify that Scopes had indeed taught evolution; the defense did not contest this testimony. The defense then tried to present testimony from a dozen evolutionary scientists and liberal theologians, all who would defend evolution as sound science. After the prosecution objected to this testimony, saying that it was irrelevant as to whether Scopes broke the law, the judge eventually sided with the prosecution. At this point, the trial had not directly addressed the supposed conflict between evolutionary science and the Bible. Darrow finally called Bryan to the stand:

> Darrow posed the well-worn questions of the village skeptic: Did Jonah live inside a whale for three days? How could Joshua lengthen the day by making the sun (rather than the earth) stand still? Where did Cain get his wife? . . . Best of all for Darrow, no good answers existed. Bryan could either affirm his belief in seemingly irrational biblical accounts, and thus expose that his opposition to teaching about evolution rested on narrow religious grounds, or concede that the Bible required interpretation. He tried both tacks at various times without appreciable success.[181]

Overall, the result of the trial was a draw. Scopes was convicted and sentenced to pay a $100 fine (later overturned on a technicality). Tennessee kept its law, but the media made the most of Bryan's statements in order to mock the backwardness of the Biblical literalists. The trial served as the subject for the 1955 play and 1960 movie *Inherit the Wind*.

The Tennessee law stayed on the books for the next four decades. After the trial, Mississippi and Arkansas both passed similar laws, and other states passed lesser restrictions. For the next three decades, many school districts and textbook publishers opted to avoid the issue by quietly dropping the subject of evolution from the classroom.

Creationism Strikes Back Again

BY THE 1960s, evolution by natural selection had been established as the fundamental integrating principle of biology, and a series of new biology textbooks—featuring evolution in a central role—was introduced to the public schools. Over half of U.S. public high schools started using these textbooks. In 1968, the U.S. Supreme Court struck down the "monkey" laws that prohibited the teaching of evolution in public schools.[182]

With evolution re-entering public schools, many fundamentalists looked for intellectual ammunition, and they turned to a recent book titled *The Genesis Flood*, by John Whitcomb and Henry Morris. Published in 1961, *The Genesis Flood* placed what it called "creation science" on an equal footing with "evolution science," as two competing scientific views. It promoted the view known as "flood geology," which held that all the fossils came from the Biblical deluge. This was largely a re-write of George McCready Price's book, *The New Geology*, using more recent scientific examples.

Organizations were founded to promote this new study. One key organization was the Institute for Creation Research in San Diego. As Henry Morris, the institute's director, made clear, their priority was not on genuine science:

> The only way we can determine the true age of the earth is for God to tell us what it is. And since He *has* told us, very plainly, in the Holy Scriptures that it is several thousand years in age, and no more, that ought to settle all basic questions of terrestrial chronology.[183]

In order to get creationism into the public schools, its supporters needed to drop the explicit references to the Bible and to start referring to "creation science" as a competing theory to "evolution science." In 1981, Arkansas and Louisiana passed "balanced treatment" laws for public schools, requiring the teaching of "evolution science" to be balanced with equal time for "creation science." Both laws were challenged almost immediately. In 1982, a federal court in Arkansas ruled the "balanced treatment" law unconstitutional (violating the Establishment Clause of the First Amendment), holding that "creation science" is religion and as such cannot be taught in public schools.

The Louisiana case was appealed up to the U.S. Supreme Court, which, in 1987, ruled that the Louisiana law was unconstitutional (violating the Establishment Clause), because its advocates had explicitly religious motives.[184] (The court focused on the *intent* of the Louisiana law.) Unlike the Arkansas ruling, this ruling left open the issue of whether the teaching of creationism as such was religious. They implicitly left open the possibility that there might be scientific evidence for creationism. This left a legal loophole for the creationists. They just had to disguise their motives better to get creationism into the public schools.

Intelligent Design: Creationism in Camouflage[185]

BY THE MID-1990S, "scientific creationism" had adapted and emerged in a new form called "Intelligent Design" or ID. The advocates of this new view, unlike those of creation science, granted the possible validity of the geological findings of an ancient earth. But like the earlier creationists, they set out to advance a theistic worldview in the schools and universities.

In the mid-1980s, several Protestant scientists wrote books arguing that entirely natural causes could not account for the origins of life, and that this fact was indisputable evidence for a God-like being.[186] Their books, being somewhat technical in content, did not reach a broad audience. Then in the 1990s, law professor Phillip Johnson described a strategy for regaining the upper hand for religion against advocates of evolution. He called this strategy "the wedge," and the first step he advocated was to attack the "ideology of scientific materialism." This strategy was on full view in his 1991 book *Darwin on Trial*, which criticized the "naturalistic" bias of science—a bias that unnecessarily limited the range of possible explanations and excluded supernatural factors.

The first popular presentation of intelligent design from a scientist was a 1996 book by microbiologist Michael Behe titled *Darwin's Black Box: The Biochemical Challenge to Evolution*. Behe argues that science is unfairly biased against non-natural explanations, and that it should be open to such explanations when entirely natural ones cannot be found.

Behe presents an idea he calls "irreducible complexity." One of Behe's favorite examples of an "irreducibly complex" system is the basic household mousetrap, which consists of five components: platform, spring, hammer, holding bar, and catch. If any of these components is missing, it is not the case that the mousetrap will catch mice less well; it will not catch mice at all. For an "irreducibly complex" system to function at all, it needs a complete set of components, and it requires a conscious designer to put them together.

Behe argues that numerous biological systems are "irreducibly complex." His favorite example is the bacterial flagellum, which resembles an outboard motor. It has a complex assembly of parts that cause a rotating shaft to propel the bacterium forward. Behe correctly observes that if you take away any of its components, the shaft will not turn at all.

Evolution can only design systems by relatively small increments, in which each stage has an increased survival-value for the organism. So all of these parts must have come into being at the same time, which seems highly improbable. So Behe argues that it must have been designed by an "intelligent agent." According to the theory of evolution by natural selection, each step (of adding a component) should add something to the survival value of the system, but, argues Behe, this is not the case.

A major flaw in Behe's argument is the assumption that, as a biological organ incrementally evolves, it always keeps performing the same function. But when a component is added, or when a small change is made, sometimes the system starts performing an entirely different function. In fact, the bacterial flagellum closely resembles the simpler structure of the "syringe" used by the *Yersinia Pestis* bacteria to inject other cells. The "syringe" structure lacks several of the components of the flagellum structure, so it doesn`t rotate at all. However, it works perfectly well at injecting host cells.

A similar type of argument to Behe's is offered by mathematician William Dembski, who proposes a statistical "explanatory filter" by which he claims that he can differentiate between necessity, chance, and design. He argues that if an event's likelihood of being naturally caused is improbable enough, then it must have a supernatural cause.

At bottom, the Intelligent Design movement is just the same old "argument from design" advocated by Reverend William Paley, repackaged using modern terminology with examples from molecular biology. Intelligent design explicitly attacks the metaphysical naturalism at the base of science. But scientific evidence cannot lead to the supernatural. The idea of the supernatural is a contradiction and an absurdity.

"Supernatural" means above or beyond nature, and "nature" is everything that exists in the universe, as viewed from a certain perspective. "Nature" is existence regarded as a lawful realm of cause and effect.

Intelligent Design on Trial

A CRITICAL TEST for the teaching of Intelligent Design in public schools came in 2005, in Dover, Pennsylvania. In order to make students aware of alternatives to evolution, the Dover Area School District Board provided a statement which ninth-grade biology teachers were required to read to their classes:

> The Pennsylvania academic standards require students to learn about Darwin's theory of evolution and to eventually take a standardized test of which evolution is a part.
>
> Because Darwin's Theory is a theory, it is still being tested as new evidence is discovered. The Theory is not a fact. Gaps in the Theory exist for which there is no evidence. A theory is defined as a well-tested explanation that unifies a broad range of observations.
>
> Intelligent design is an explanation of the origin of life that differs from Darwin's view. The reference book, *Of Pandas and People*, is available for students to see if they would like to explore this view in an effort to gain an understanding of what intelligent design actually involves. As is true with any theory, students are encouraged to keep an open mind.[187]

Sixty copies of *Of Pandas and People* were donated to the school library by one of the board members, using funds raised from his church congregation. *Of Pandas and People: The Central Question of Biological Origins* was initially published in 1989 by Dean Kenyon and Percival Davis. It was intended to supplement high school biology textbooks. *Of Pandas and People* defines "intelligent design" as a frame of reference that "locates the origin of new organisms in an immaterial cause: in a blue-print, a plan, a pattern, devised by an intelligent agent."[188] The book discusses six separate puzzling issues in biology, such as the origin of life on earth. For each topic, the book presents the Darwinian explanation and contrasts it to the "intelligent design" explanation, which it argues is superior.

The biology teachers in the district all refused to read the official statement, so an administrator came into their classes to read it in their stead. Eleven parents appealed to the ACLU, which got them a team of lawyers for their civil suit against the school board. The resulting trial began in Harrisburg, Pennsylvania in September 2005. The case, like the earlier creation-science trials, hinged on whether the teaching of intelligent-design theory, as shown in *Of Pandas and People*, was an example of the teaching of religion, and so violated the First Amendment.

The judge was the conservative John E. Jones III, who had been appointed by President Bush in 2002, so members of the board were cautiously optimistic about the outcome. However, at the end of the six-week trial, Judge Jones' final verdict was severely critical of the Dover school board, describing its actions as "breathtaking inanity." He ruled that Intelligent Design failed "to meet the essential ground rules that limit science to testable, natural explanations," and therefore it was "not science." Jones concluded that "it is unconstitutional to teach Intelligent Design as an alternative to evolution in a public school science classroom."[189] This ruling applied only to the middle district of Pennsylvania, but its influence extended nationwide.

* * *

In this chapter and the previous one, we have seen how science destroyed any literal interpretation of the book of Genesis and the origin of man. With the story of Adam and Eve and the Fall of Man seen now as a mere fable, the validity of the entire Bible was severely undermined.

Darwin's entirely natural theory of evolution pulled the rug out from under the argument from design, by showing how the complexity of living things can arise from a completely natural process.

A book is a product of a conscious mind. Elevating a book above the evidence of one's own observations and reasoning is the elevation of consciousness over existence. The primacy of the Bible *is* the *primacy of consciousness*. The advance of Darwin's theory of evolution involved a severe downgrade of the Biblical view of man's origin, based on ful-ly-validated observation-based theory. The acceptance of Darwinism (and later neo-Darwinism) asserted the *primacy of existence*.

CHAPTER TWELVE

Kant Enables
the Religionists

F ROM AQUINAS TO the Renaissance, to the Scientific Revolution, to the Enlightenment, we have seen an ever-expanding acceptance of the *primacy of existence*, and a corresponding demotion of religion. But the trend did not continue long beyond this, because of a new form of the *primacy of consciousness*—the *social* primacy of consciousness—which became an enabler for religion.

The creator of this new approach was the Prussian philosopher Immanuel Kant (1724–1804). Kant is the most influential philosopher of the modern world, and he profoundly shaped the way that we think about science and its relation to religion. Kant's philosophy was motivated by the need to answer David Hume, who had unhappily arrived at the conclusion that we cannot know anything at all.

Hume's Problem

THE PHILOSOPHER DAVID Hume had started with the empiricist assumption that all our knowledge begins with discrete sensations, such as bits of color, light, and sound. Then the problem is to explain how we can assemble these together, in order to form all of our knowledge. Unable to do this, Hume concluded that we can't know anything about the external world or even that we have selves.

Consider one key aspect of our experience: causality. We see (and rely on) cause-and-effect relationships throughout our lives. Hume noticed that when we see one event "cause" another, we cannot really see the cause-and-effect relation itself.

Consider using a pin to pop a balloon. Whenever we prick a normal balloon with a sharp pin, the balloon pops with a loud bang. It is obvious that the prick *causes* the balloon to pop. But how do we know this? How do we know that this is an instance of causality?

Hume would say that we can see and feel the pin pricking the balloon, and we can hear and see the balloon popping. But we cannot see the "causality." We can just see that the two events happen at the same time and in the same place. But we cannot see the necessary connection between the cause and the effect. We are simply conditioned to seeing events as necessarily connected, but we don't know that they are. For all we know, the next time we prick a balloon, it might not pop at all; it might do something completely different.

Hume concluded that we have no good reason to think that causality will continue to hold in the future. We can't even know that the sun will rise tomorrow. Hume was not happy with this conclusion, but he did not see anything wrong with his line of argument.

Kant's Philosophic System

IN WORKING TO solve Hume's problem, Kant ended up creating a system of philosophy that, Kant claimed, saved knowledge and objectivity. His starting point was the claim that our preconscious minds are complex synthesizing mechanisms—which process the data that they get from reality in many different ways before the data reach conscious awareness. The world of our experience is the result of this processing, and it is all that we are aware of.

Kant had, in essence, split reality into two parts: the noumenal world, which consists of "things-in-themselves," and the phenomenal world, which consists of our conscious experiences. Because of the processing of our mental machinery, we are forever cut off from "things-in-themselves." They are unknowable.

In response to Hume's problem with causality, Kant answers that, while we cannot know "things-in-themselves," we can count on our experiences to exhibit causality. This is because the output of our mind's processing includes causality. The mind does not *discover* law and order in the world; it *creates* law and order through its preconscious synthesizing activities.[190]

Kant provided an elaborate chain of arguments to support his claims about the specific ways that our minds construct our experience. At the sensory/perceptual level, our minds introduce space and time. At the conceptual level, our minds shape their input in twelve different ways; these are Kant's twelve categories. These categories include, among other things, causality and substance (entity).

Kant correctly claimed that he had reversed the direction of philosophy, in a way comparable to the Copernican Revolution in astronomy. Before Kant, all philosophers accepted that there is a reality out there, there is a mind in here, and the mind needs to conform to reality. But with Kant's reversal, *objects must conform to our minds*. Reason cannot know reality, but Kant did not consider this a problem; we can reconceive what reason *is*, so that it does not depend on reality.

Kant did defend the idea of "objectivity," sharply distinguishing between the subjective and the objective. But, as with knowledge itself, he reconceived the meaning of these concepts. For Kant, "objective" refers to those aspects of human experience that are universal, because they are part of the structure of the human mind. Thus Kant's "objective" amounts to collective subjectivity.

Kant has often been considered an advocate and defender of reason. However, as philosopher Stephen Hicks points out:

> the fundamental question of reason is its relationship to reality. Is reason capable of knowing reality—or is it not? . . . Kant was crystal clear about his answer. Reality—real, noumenal reality— is forever closed off to reason, and reason is limited to awareness and understanding of its own subjective products.[191]

This connects to an important part of Kant's motivation, as Hicks explains:

> One purpose of the *Critique* . . . was to limit severely the scope of reason. By closing noumenal reality off to reason, all rational arguments against the existence of God could be dismissed. If reason could be shown to be limited to the merely phenomenal realm, then the noumenal realm—the realm of religion—would be off limits to reason, and those arguing [using reason] against religion could be told to be quiet and go away.[192]

As Kant tells us in the Second Preface to his *Critique of Pure Reason* (1781), "I have therefore found it necessary to deny *knowledge,* in order to make room for *faith.*"[193]

Kant argued that one could obtain a lead into the noumenal world through the field of ethics. Kant's ethical views hinge on the idea that consequences cannot be the means to justify ethics. Moral laws must declare certain actions as right in themselves, regardless of anybody's goals, purposes, or consequences.

Kant's moral views, he claimed, are grounds to conclude that God, free will, and immortality must exist, but he could not find these in the phenomenal world.[194] He concluded that these must exist in the noumenal world; they cannot be known through reason, but they can be grasped through faith.

Kant's Descendants

IT IS VIRTUALLY impossible to find a significant philosopher of the last two centuries who does not accept at least some of Kant's basic framework. Let's consider a few of the major philosophers of science of the twentieth century.

Sir Karl Popper, the alleged defender of science, accepted the Kantian idea that reality is impossibly cut off from our ideas. Instead, he held, science must be based on inter-subjective "observation statements" on which many can agree. A scientist's own observations cannot put him in contact with reality, but if they are stated publicly and can be agreed on by others, then that is what counts. This view places the group's belief, not reality, at the base of science. Based on a process of trial-and-error, with conjectures being proposed and then refuted by observation statements (as discussed in Chapter Four), Popper held that we arrive at theories that get, in some sense, closer and closer to the truth. However, Popper held that the current theory will almost certainly be proven false by future observations. (And anyway, "true" here refers to Popper's inter-subjective truth.)[195]

According to philosopher David Stove:

> The truth of any scientific theory or law-statement, [Popper] constantly says, is exactly as improbable, both *a priori* and in relation to any possible evidence, as the truth of a self-contradictory proposition; or to put the matter in plain English (as Popper does not), it is impossible.[196]

The skepticism in Popper became more explicit with his student Paul Feyerabend, who concluded that there are no valid general methods of science. Feyerabend held that western science has been highly overrated—that it has been unjustly placed on a pedestal above non-western traditions such as voodoo. Calling his view an "anarchistic theory of knowledge," he rejected theories as such, and concluded that "anything goes."[197]

Feyerabend's extreme skepticism was too blatant for most historians of science. They were drawn instead to Thomas Kuhn, whose attack on reason was far more subtle. When Kuhn published *The Structure of Scientific Revolutions* in 1962, it was a landmark in the history and philosophy of science, and its impact on the field can hardly be overstated. Kuhn argued that science periodically undergoes "paradigm shifts," in which one worldview is rejected and a new one takes its place. A prominent example of such a shift was the overthrow of the Ptolemaic geocentric cosmology in favor of the Copernican heliocentric cosmology. Kuhn held that competing paradigms are "incommensurable"—that is, they cannot be rationally compared. Ultimately there are no rational grounds for choosing one over another. Non-rational factors must always be examined to understand why, in history, a particular shift took place. One such factor is faith: "The man who embraces a new paradigm at an early stage must often do so in defiance of the evidence provided by problem-solving. He must, that is, have faith that the new paradigm will succeed. . . . A decision of that kind can only be made on faith."[198]

Kuhn's book completely removed the concepts of *truth* and *objectivity* from the history of science. From this point onward, historians of science increasingly considered the "truth" of a scientific theory as utterly irrelevant to any discussion of its place in history. When they examined a clash between two opposing views, the fact that one might be based, ultimately, not on reason but on faith, meant virtually nothing.

Today's postmodernist movement—with its total skepticism and collectivism—also traces its roots back to Kant. This is well documented in Stephen Hicks's *Explaining Postmodernism: Skepticism and Socialism from Rousseau to Foucault.*

Impact on "Science and Religion" Studies

MOST HISTORIANS (like most people) do not make the premises of their philosophy clear to themselves. But the premises operate nonetheless when they practice their craft. When a historian influenced by Kantian-based philosophy confronts an issue of reason and faith in history, he is bound to blur the difference between the two—the distinction between reason and faith, to him, is virtually meaningless.

The post-Kantian skepticism, supported in part by Popper, Feyerabend, and Kuhn, also entailed a distrust of broad abstractions such as "science" and "religion," which many historians of science claimed to be impossible to define. For example, does "religion" refer primarily to a set of beliefs, a method of coming to conclusions, a set of practices, or a type of institution? Questions such as these are often considered unanswerable.

Without a firm grasp of such broad abstractions, historians necessarily must be concrete-bound. They will observe a scientist who is a monk and conclude, without any reference to broader issues, that this demonstrates that science and religion are compatible.

An example of the concrete-bound approach can be seen in the reasoning of historian John Heilbron, who described how a large number of seventeenth-century Jesuits studied and taught about static electricity. He concluded that "The single most important contributor to the support of the study of physics in the seventeenth century was the Catholic Church."[199] This was the same century as the Church's condemnation of Galileo!

As historians of science looked into the relationship between science and religion in history, they found evidence of a complex variety of interactions. Sometimes religion seemed to hinder science, but often there appeared to be no conflict at all. Most scientists in history were religious, and many were more religious than average in their societies, yet their religion did not seem to impede their work. Much scientific work was performed by clerics, such as the Polish canon Nicolaus Copernicus, the Augustinian abbot Gregor Mendel, and the numerous Jesuits who studied static electricity.

Another connection between science and religion is the discipline known as "natural theology" in which scientists seemed to intimately link their religion and their science. Figures such as the chemist Robert Boyle and the naturalist John Ray saw their scientific work as bringing them closer to God through knowledge of God's creation. These are the kinds of examples that a concrete-bound historian will regard as conclusive evidence for the compatibility of science and religion.

History as a discipline relies on views of the nature of reality, knowledge, and values—i.e., it relies on philosophy. But philosophers have been relentlessly attacking reason—which is the base of science. When philosophers can no longer tell the difference between reason and faith, it is not surprising that our historians cannot either, and that they cannot see any conflict between science and religion.

An Answer to Kant

KANT'S ARGUMENT, in essence, amounts to the following: Because human consciousness has a specific process for perceiving reality, then it cannot perceive reality in itself. But the same argument would apply to the consciousness of an ant, an elephant, or an alien from another galaxy. According to this argument, no consciousness, not even God's (if it had an identity), could perceive reality "as it really is." Rand observes that:

> All knowledge *is* processed knowledge—whether on the sensory, perceptual or conceptual level. An "unprocessed" knowledge would be a knowledge acquired without means of cognition. Consciousness . . . is not a passive state, but an active process. And more: the satisfaction of every need of a living organism requires an act of *processing* by that organism, be it the need of air, of food, or of knowledge.[200]

Any type of consciousness must have a specific *identity* and must do some processing. Each perceives in a form specific to that species. But that does not invalidate the consciousness; that does not mean it is not conscious of reality.

What would it mean for a consciousness to grasp reality "as it really is" as opposed to how reality "appears" to it? It would mean that the consciousness performs no processing, and has no means of being conscious. This view is incoherent. There is no such thing as "reality as it really isn't." In the words of Harry Binswanger, "Every appearance is the appearance of reality."[201]

We can be mistaken in our interpretation of our perceptions (such as in the "stick bent in water" illusion), but it is ultimately our perceptions that allow us to correct these mistakes. (For a detailed discussion of representationalism and two other conceptions of consciousness, see Chapter 2 of Harry Binswanger's book *How We Know*.)

While Kant's philosophy is not a religion, it does have one interesting commonality with religion. It is a form of the *primacy of consciousness*—based not on a divine mind, but on a collective of human minds. Most importantly, Kant's philosophy enabled religionists to hold onto their faith.

CHAPTER THIRTEEN

Reason, Morality, and Purpose

O NE CRUCIAL REASON for the continued influence of religion today is that it is considered the only source of objective moral guidance. Religion has pre-empted the field of morality, including the way people think about meaning and purpose in their lives. Even prominent atheists and secular "skeptical" organizations share the premises of religious ethics.

Morality needs a foundation in reason, not revelation or subjective feelings. In order to find a reason-based approach to morality, let us first turn to the greatest philosopher of the culture that created science.

The Ethics of Aristotle

ARISTOTLE IS ARGUABLY the first philosopher to create a fully secular theory of morality. For Aristotle's teacher, Plato, morality comes from a God-like Form of the Good, but for Aristotle, morality comes from the observable facts of human nature.

Like most of the Greeks around him, Aristotle saw ethics or morality not centered on self-sacrifice, as in Christianity, but as the means to a successful life. For Aristotle, morality is about living one's own life to the fullest, being a "great-souled man" and being justly proud of one's accomplishments. He argues that, if one takes the phrase "lover of self" in a proper sense, "the good man should be a lover of self (for he will both himself profit by doing noble acts, and will benefit his fellows)."[202]

Aristotle claimed that the proper end of one's life is *Eudaimonia,* which is often translated as "happiness," though it means something closer to "living a full, flourishing human life." He held that this could generally be achieved (assuming no serious misfortune) by living according to a set of virtues, including courage, justice, temperance, and pride.

Virtues do not give us a set of rigid out-of-context rules or commandments (such as "do not lie"), which we must follow; rather, they are guides—principles of action to be applied appropriately in any given situation.

Unfortunately, Aristotle's ethical system was not historically influential, in part due to its own deficiencies. His theory of the golden mean presented each virtue as the mean on a scale between extremes. For example, when it comes to facing danger, having too much fear is the vice of cowardice; having too little fear is the vice of foolhardiness; having just the right amount of fear is the virtue of courage. How is one to determine the "right amount" for each virtue? Aristotle did not have a good answer. He seems to imply that one should observe and emulate those who are regarded as great men. Ultimately, Aristotle's defense of his virtues seems to be circular.

Life as the Source and Standard of Value

THE CONCEPTS "good," "evil," "right," and "wrong," at the root of morality, are concepts of *value*. But is this type of concept "an arbitrary human invention, unrelated to, underived from and unsupported by any facts of reality?"[203] If not, what facts of reality give rise to the need for such concepts? This was the question asked by Ayn Rand. [204]

Rand answered that it is the phenomenon of *life* that leads to "good," "evil," "right," "wrong," and all the other concepts of value. She characterized life as "a process of self-sustaining, goal-directed action." Every aspect of a living organism is goal-directed and purposeful (although generally not at a conscious level). A plant turns its leaves toward the sun in order to get light; it probes its roots into the ground to get nutrition. A squirrel gathers nuts in the fall in order to sustain itself during the winter. And it does all of the other things it does in order to sustain its life.[205]

Rand observed that life is conditional; living things must take certain actions, or they go out of existence—they die. This led Rand to the insight that life is the root of all value. All values have a "to whom" and a "for what." The "to whom" is the organism itself. The "for what" is the organism's life.[206]

With plants and many animals, it is often easy to see when a living organism is living successfully, or flourishing. It is healthy; it is taking all the actions needed to sustain its life, which is its ultimate purpose. In contrast, sometimes we see a plant or animal which is barely surviving; it is alive but not really *alive*.

For human beings, the same issue of "flourishing" exists. As Sam Harris correctly points out in his book *The Moral Landscape*, "the difference between a healthy person and a dead one is about as clear and consequential a distinction as we ever make in science. The differences between the heights of human fulfillment and the depths of human misery are no less clear."[207] (However, as we will see, Harris comes to very different ethical conclusions than Rand does.)

The complexity of human well-being puts it on a different order than that of the other animals. Humans have a distinctly volitional and conceptual consciousness. They must continuously make choices, where every action has possible long-range consequences. Rand argues that, in order to flourish, human beings need a code of morality to guide their choices and actions. This is what morality is for.

Rand argues that the standard of moral value is "man's life," or "that which is required for man's survival *qua* man."[208] For Rand this means a full, flourishing life as a rational animal, centered on three fundamental values: *reason, purpose*, and *self-esteem*.

Reason has been discussed in earlier chapters. Here it is important to note that reason is our basic means of survival, but it is not automatic; it must be exercised by choice and the exertion of effort. Holding reason as a fundamental value means choosing to place it at the center of one's life.

Holding *purpose* as a fundamental value means organizing one's life around wide goals such as a productive career, as we will see in the next section.

The last of these fundamental values, *self-esteem*, is, in Rand's words, "[man's] inviolate certainty that his mind is competent to think and his person is worthy of happiness, which means: is worthy of living."[209] Self-esteem must be earned by consistently acting in accordance with one's values.

The moral principles necessary for the achievement of these values may be summarized by the seven virtues that Rand identifies: rationality, independence, productiveness, integrity, justice, honesty, and pride. The central virtue of this list, which underlies all the others, is *rationality*— the commitment to the full use of *reason* as a guide to one's thoughts and actions. Rand conceives each of the other virtues as a form of recognizing facts of reality and acting accordingly. For example, she writes:

> Honesty is the recognition of the fact that the unreal is unreal and can have no value, that neither love nor fame nor cash is a value if obtained by fraud—that an attempt to gain a value by deceiving the mind of others is an act of raising your victims to a position higher than reality, where you become a pawn to their blindness, . . .[210]

For an in-depth discussion of all of these virtues, including issues such as benevolence and charity (which are entirely compatible with Rand's ethics), see Tara Smith's *Ayn Rand's Normative Ethics: The Virtuous Egoist*.

Purpose as Self-Generated

It has often been observed that a sense of purpose is an important component of a happy life. The person who sets no wide goals in his life, has no important values, and lives only moment to moment, will not be happy.

In the religious view, meaning and purpose are things that come from outside oneself. We are all supposedly actors in God's play, and once you figure out your role in this play, you can say, "I found my purpose in life" or "I found the meaning of my life." According to Pastor Rick Warren, "Without God, life has no purpose, and without purpose, life has no meaning."[211]

But the truth is that purpose and meaning do not come from outside; they come from within oneself. Purposes must be chosen according to one's own personal values, and they must be chosen rationally, by an objective standard.

An important type of purpose is one's productive work, or career. There is a virtually unlimited number of types of productive work one might pursue. One might work as a teacher, an engineer, a lawyer, a farmer, a fireman, a businessman, or any other legitimate profession (as opposed to being a bank robber or panhandler). Parenting itself may be seen as a productive purpose. One's productive purpose may change several times over the course of one's lifetime.

The acts of defining one's central purposes and of acting on them over time are what give one's life "meaning." A meaningful life is one that is filled with the pursuit of important values—values that are personally important to oneself.

* * *

Rand contrasted her morality with the culturally-dominant morality of *altruism*, which, contrary to popular understanding, does not mean benevolence or "being nice to other people." Coined by philosopher Auguste Comte in the nineteenth century from the Latin root for "other," the term "altruism" means "other-ism." It means that you have a *duty* to sacrifice your own values for the good of other people.[212]

Although the term is historically recent, the idea of altruism is fairly widespread in history, appearing in some form in virtually all religions. Consider the Jesus of Christianity: he is seen as the ultimate altruist, sacrificing himself completely for the good of humanity.

The New Athiests and the Morality of Christianity

THE ATROCITIES OF 9/11 provided a horrific demonstration of the destructiveness of religion. A handful of atheist writers, sometimes called "the new atheists," reacted by writing books about the harmfulness of religion. These writers included Christopher Hitchens, Richard Dawkins, and Sam Harris, among others. The new atheists have been courageous and articulate in their attacks on religion. But their own views on the foundation of morality have been, at best, not particularly coherent, and at worst, taken directly from religion.[213]

Christopher Hitchens says that there is an "innate conscience" possessed by "everybody but the psychopath," which tells us what is right and wrong.[214] But this ignores the variety of ethical norms of individuals and societies throughout history. As just one example, the practice of slavery was considered morally proper through much (if not most) of human history. Does that make it acceptable? How does this "innate conscience" allow us to *just know* what is right and wrong? Hitchens has no answer.

Richard Dawkins talks about a "moral Zeitgeist" that is a general consensus within a society about what is right and wrong.[215] This "moral Zeitgeist" is always changing, as people change their ethical views over time. But this amounts to social subjectivism.

Sam Harris seems to be more serious in his attempt to provide a scientific basis for morality. He claims to subscribe to "ethical realism"—the view that, "in ethics, as in physics, there are truths waiting to be *discovered*—and thus we can be right and wrong in our beliefs about them."[216] Harris sees the fundamental fact behind ethics to be that conscious organisms can feel pleasure or pain.

Harris grounds his ethics in his "intuitions." These tell him that the standard of the good is happiness, and the standard of the evil is suffering. However, he does not mean that one should pursue one's own happiness and avoid one's own suffering. The standard of value is the total pleasure and pain *of every conscious organism in the world*. Given the many billions of such beings, the value of one's own happiness shrinks effectively to zero. (This is essentially similar to the theory of utilitarianism developed by Jeremy Bentham, in which the standard of good is often summarized as "the greatest good for the greatest number.")

The resulting moral ideal becomes virtually identical to that of Christianity: self-sacrifice for others.

There is one essential commonality between these views of the foundation of morality. Hitchens's "innate conscience," Dawkins's "moral Zeitgeist," and Harris's "intuitions" all amount to the consciousness of individuals or groups of people. These are all examples of the *primacy of consciousness*. In contrast, Rand establishes the foundation of her morality in basic observable facts of reality, notably in our nature as living beings who survive by choosing to use our conceptual faculty. Rand's view is based on the *primacy of existence*.

CHAPTER FOURTEEN

Science, Technology, and Our Environment

> Man is a singular creature. He has a set of gifts which make him
> unique among all the animals: so that, unlike them, he is not a
> figure in the landscape — he is a shaper of the landscape. In body
> and in mind he is the explorer of nature, the ubiquitous animal,
> who did not find but has made his home in every continent.[217]

So begins Jacob Bronowski's book *The Ascent of Man.*

Compared to the other animals, Bronowski notes, man possesses an
extremely crude survival kit; he is not particularly strong, or fast, or of
keen sight, smell, or hearing. But he possesses one extraordinary tool—a
mind that uses reason—which allows him to project far into the future
or the past, to make plans for action, and to shape his environment in
accordance with his needs. He can act on a grand scale, in a way that
makes him different in kind from all other animals.

Humans have never really lived in "harmony" with their environment; they have always shaped it, to some extent. Early humans, in
hunter-gatherer societies, probably hunted large animals to extinction.[218] The first dramatic landmark of human history, and the key to the
development of civilization, is known as the "agricultural revolution" or
the "neolithic revolution." This occurred roughly 12,000 years ago, at
the end of the last ice age, when humans began cultivating plants and
domesticating animals. This led to the widespread clearing of land and
the reshaping of ecosystems.

The development of technology progressed in fits and starts, often
in ways not directly connected to scientific activity. The small machines

and mechanisms developed by ancient Greeks, such as Archimedes, were seen as toys and curiosities—not actually useful for real work. The Roman architectural advances such as the arch and concrete did not emerge out of scientific studies. Over the course of the Middle Ages, a handful of simple technologies gradually came into widespread use; these included the waterwheel mill and the iron plow.

The Industrial Revolution

IN THE HISTORY of technology, the most important single landmark is the Industrial Revolution, which began in eighteenth-century England. The dominant cause was the culture of the Enlightenment. This was a culture of reason, science, worldliness, and relative political/economic freedom (capitalism). People were free to think, experiment, invent, and to invest in new types of ventures. As a result, they created a revolution in productivity.

James Watt's steam engine, which he developed in the 1770s, was perhaps the most important invention of the Industrial Revolution. It created a practical way to use heat (from burning wood or coal) to generate regular rotation of an axle. This could then drive a wide variety of different machinery, leading to steam-driven factories, steamboats, and railroads.

The factory system allowed the inexpensive creation of an abundance of consumer products. Consider the garment industry: Between 1784 and 1860, the price of cotton cloth decreased by 99%, allowing hundreds of millions of people to dress comfortably and cleanly.[219]

The industrial revolution spread out from England to America, Europe, and then the rest of the world, leading to a radical transformation of human life. The internal combustion engine led to practical automobiles; together with trains, these made it possible for the nonwealthy to easily travel long distances for the first time in history. The development of electricity enabled the safe and practical lighting of homes, as well as countless labor-saving devices.

As a result of all this technology, in an atmosphere of freedom and capitalism, there was more and more development of improved products and more efficient methods of production. We have had more and better: food, clothing, shelter, medical care, countless conveniences of life, and a significant reduction in human fatigue and exhaustion while extending our lifespans. These have led to the end of famines and plagues; they have eliminated the terrible diseases of cholera, diphtheria, smallpox, tuberculosis, and typhoid fever. Industrial civilization has actually *ended* famine; it has done this by creating an extraordinary abundance and variety of food and created the storage and distribution systems to bring it to everyone.

As always in history, new technologies often led to unforeseen problems; but in general, the benefits far outweighed the problems. The early factories were not pleasant places to work by modern standards, but their workers (which included children) came from backbreaking work in the fields, fourteen-hour days, and the risks that their crops would be lost to drought or flood.

The economic system of the present-day industrial world is a truly remarkable achievement. The industrialization that it spawned is still growing and transforming the world. This transformation has been overwhelmingly positive for human well-being. We should not take for granted the freedom, science, and technology that led to this.

This progress in human well-being has been thoroughly documented by Steven Pinker in his 2018 book *Enlightenment Now: The Case for Reason, Science, Humanism, and Progress*. Consider life expectancy and child mortality. Starting in the nineteenth century, global life expectancy began to rise, picked up speed in the twentieth century, and is continuing with no slowdown. Through most of human history, child mortality was at appalling levels: one-fifth of children died in their first year, and almost half died before reaching adulthood. Then, starting in the nineteenth century, rates of childhood mortality started a rapid decline, which continues to this day.

Pinker documents similar trends in many other measures of well-being, including deaths from infectious disease and accidents, and amounts of famine, malnutrition, and pollution. He also documents dramatic and continuing positive worldwide trends in wealth, education, and safety.

Dramatic progress is even apparent in the statistics of *climate-related* deaths (i.e., fatalities due to heat waves, cold spells, droughts, floods, and storms). From the 1920s to 2010, the total mortality rate from these events, globally, declined by 98%.[220] This is not so surprising if we consider that global poverty levels have fallen dramatically, and that a nation's wealth makes a considerable difference in its ability to build effective infrastructure that protects people from dangerous climatic events.

An important part of our environment is the quality of our drinking water, and this quality depends on economic development. Water quality depends on chemical purification plants, pipelines, and pumping stations. Air quality in cities has always been poorer than in the open country, but today it is much better than in the past. Before modern industry, public streets in cities were open sewers, and horse manure was everywhere.

Environmentalism and the Quasi-Religion of Greenism

OF ALL THE polarizing issues in the world today, there is perhaps none more so than environmentalism, especially in the United States. Political conservatives denounce what they call the "global warming hoax," and liberals accuse "climate change deniers" of crimes against humanity. Many politicians and activists talk about the "existential threat" posed by climate change and push for government action to quickly phase out fossil fuels (which currently provide about 85% of the world's energy).[221] However, much of the public is not on board with this; recent Gallup polls show that only 3–4% of Americans specify environment as a top concern.[222]

The modern environmental movement took shape in the 1960s. It emerged out of a combination of two factors: (1) a set of recent scientific discoveries about climate and ecology; (2) a romantic philosophic movement having a reverence for untouched nature. Natural phenomena revealed by the science included:

- The approximate trajectories of climate changes that the earth has undergone in the past
- The carbon cycle, in which carbon flows between the atmosphere, ocean, and land
- The types of climate feedbacks that can affect global warming and cooling trends

The romantic philosophic movement had its roots in eighteenth and nineteenth-century thinkers such as J. J. Rousseau and Friedrich Hegel (a descendant of Kant), who reacted strongly against the Enlightenment's commitment to reason as the means of knowing reality and guiding action.

Steven Pinker points out that "the mainstream environmental movement latched onto a quasi-religious ideology," which he calls "greenism." This ideology "can be found in the manifestos of activists as diverse as Al Gore, the Unabomber, and Pope Francis."[223] Greenism has a "pathological fear" of industrial civilization, science, and technology; as such, it is the most reactionary movement in the history of the world.

According to this ideology, "Unless we repent our sins by degrowth, deindustrialization, and a rejection of the false gods of science, technology, and progress, humanity will face a ghastly reckoning in an environmental Judgment Day." Pinker continues: "As with many apocalyptic movements, greenism is laced with misanthropy, including an indifference to starvation, an indulgence in ghoulish fantasies of a depopulated planet, and Nazi-like comparisons of human beings to vermin, pathogens, and cancer."

What is *religious* about this ideology? Author Michael Crichton gives us some clues, arguing that it is "a perfect 21st century remapping of traditional Judeo-Christian beliefs and myths:"

> There's an initial Eden, a paradise, a state of grace and unity with nature, there's a fall from grace into a state of pollution as a result of eating from the tree of knowledge, and as a result of our actions there is a judgment day coming for us all. We are all energy sinners, doomed to die, unless we seek salvation, which is now called sustainability. Sustainability is salvation in the church

of the environment. Just as organic food is its communion, that pesticide-free wafer that the right people with the right beliefs, imbibe.[224]

Similar comparisons of the green ideology to religion have been made by numerous accomplished scientists, including physicist Freeman Dyson[225] and ecologist James Lovelock (author of the famous Gaia Hypothesis).[226]

The similarities between greenism and religion include the following: (1) a standard of value outside of human life, (2) recurring predictions of doom, (3) authoritarianism, and (4) a focus on rituals.

1. A Standard of Value Outside of Human Life

Religion has always placed the ultimate source of moral value to be outside of the lives of actual human beings. This source has typically been God and a supernatural realm. Greenism agrees with conventional religion that the source of moral value must be outside of mankind, but it holds that this source is pristine nature—nature "untainted" by human influence. Many people consider wilderness to be intrinsically good, apart from its value to us. By this standard, humans are the enemy, as made explicit by many environmentalists.

John Muir, a founder of the mainstream Sierra Club, wrote that "I have precious little sympathy for the selfish propriety of civilized man, and if a war of races should occur between the wild beasts and Lord Man, I would be tempted to sympathize with the bears."[227]

David Foreman, founder of Earth First, stated that "Wilderness has a right to exist for its own sake, and for the sake of the diversity of the life forms it shelters; we shouldn't have to justify the existence of a wilderness area by saying, 'Well, it protects the watershed, and it's a nice place to backpack and hunt, and it's pretty.'"[228]

Biologist David Graber says he is one of those who:

> value wilderness for its own sake, not for what value it confers upon mankind.... We are not interested in the utility of a particular species, or free-flowing river, or ecosystem, to mankind. They

have intrinsic value, more value—to me—than another human body, or a billion of them. Human happiness, and certainly human fecundity, are not as important as a wild and healthy planet. . . . It is cosmically unlikely that the developed world will choose to end its orgy of fossil-energy consumption, and the Third World its suicidal consumption of landscape. Until such time as Homo sapiens should decide to rejoin nature, some of us can only hope for the right virus to come along.[229]

In December 2018, *The New York Times* published an op-ed piece by philosophy professor Todd May titled "Would Human Extinction be a Tragedy?" Here he argues that, in some sense, the extinction of humanity would be a "tragedy," but considering the damage that we do to the earth, it also "might just be a good thing."[230]

2. Predictions of Imminent Doom

Growing up in Chappaqua, NY, in the 1970s, I remember being told by my teachers that scientists were predicting catastrophic global cooling caused by industrial activity.[231] Within a few years, the scientific and media concern shifted to global warming along with related population and resource catastrophes. We have now had five decades of such predictions, of which here is a small sample:

- From bestseller *The Population Bomb* (1975) by ecology professor Paul Ehrlich: "The battle to feed all of humanity is over. In the 1970s and 1980s hundreds of millions of people will starve to death. . . . At this late date nothing can prevent a substantial increase in the world death rate."[232] Ehrlich also wrote in 1970: "If I were a gambler, I would take even money that England will not exist in the year 2000."[233]

- In 1970, Sen. Gaylord Nelson, D-Wisc.—often considered the "father of Earth Day"—cited the secretary of the Smithsonian, Dr. S. Dillon Ripley, who "believes that in 25 years, somewhere between 75 and 80 percent of all the species of living animals will be extinct."[234]

- In 1972, a prestigious study called *The Limits to Growth* (by the Club of Rome) predicted the exhaustion of almost every industrial resource, including aluminum (by 2003), copper (by 1993), natural gas (by 1994), and petroleum (by 1992).[235]

- In 1982, Mostafa Tolba, executive director of the UN Environment Program, warned: "By the turn of the century, an environmental catastrophe will witness devastation as complete, as irreversible, as any nuclear holocaust."[236]

- Environmentalist Bill McKibben wrote in 1989 that "the choice of doing nothing—of continuing to burn ever more oil and coal—is not a choice, in other words. It will lead us, if not straight to hell, straight to a place with a similar temperature." Also: "A few more decades of ungoverned fossil-fuel use and we burn up, to put it bluntly."[237]

- John Holdren (science advisor to Barack Obama) claimed, according to Paul Ehrlich in 1986: "As University of California physicist John Holdren has said, it is possible that carbon-dioxide climate-induced famines could kill as many as a billion people before the year 2020."[238]

Of course, none of these catastrophes came to pass. They were all not just wrong, but *spectacularly* wrong. As energy expert Alex Epstein notes: "Instead of following the leading advice and restricting the use of fossil fuels, people around the world nearly doubled their use of fossil fuels—which allegedly should have led to an epic disaster. Instead [due to the resulting availability of cheap, reliable energy] it led to an epic improvement in human life across the board."[239]

Hollywood has found much inspiration from these doomsday scenarios, turning out innumerable science-fiction movies with the same theme: "The disaster that man brings on himself when he tampers with Mother Nature." The modern media and politicians also crave disaster stories, which can be effective in getting the public's attention.

What is most notable is that, while new predictions of apocalypse keep coming out, nobody circles back and asks why the earlier predictions

were so wrong. Instead, our thought-leaders count on the previous predictions being forgotten, or they excuse them on the grounds that wild
exaggeration is necessary to motivate political action. But the terribly
wrong predictions are symptoms of an anti-progress ideology.

3. Authoritarianism

Within the religion of Catholicism, there are anointed authorities—
priests—whom we are expected to follow. In the modern world, we are
supposed to accept the conclusions of a manufactured "consensus" of
scientists, instead of demanding clear explanations of what scientists
know and don't know, and how certain they can be of their claims.

Alex Epstein has identified two different approaches one can have
to experts. One can treat them as *authorities* to be followed and not to
be questioned. Or one can treat them as *guides* who can help us understand the issues in a given field—guides from whom we should expect
clear explanations, and who should be open to a wide range of challeng-
ing questions.[240]

In medicine, we go to experts (doctors) and we generally trust their
opinions because of the good reputation of modern medicine, which has,
by its record, clearly earned its reputation. But even then, you should
not trust your doctor blindly; when the stakes are high, you should get
second and third opinions, and then make sure you fully understand
your options.

Has climate science earned our trust regarding the prediction of
future climate changes? The Intergovernmental Panel on Climate
Change's global climate models have consistently predicted greater
warming than we have observed. Climate scientists such as Patrick
Michaels have argued that the models have exhibited systematic prediction failures showing that their sensitivity to CO_2 level was too high.[241]

The term "climate denier" or "denialist" is now regularly used to
refer to *any* climate thinker who sees likely climate changes as less
than apocalyptic. However, most of these thinkers fully accept that climate is changing and that human activity is contributing to it; some

of them refer to themselves as "lukewarmers."[242] "Denier" is a smear term used to shut down debate; an example of the logical fallacy called "Argument from Intimidation," it does not belong in any rational discussion.[243] (In addition to "denier," there are other widely-used terms that do not have clear definitions, and which seem to be purposefully vague. These include "sustainability," which organizations across the world are signing up for, although no one can explain exactly what it means.[244] "Climate change" itself is a deliberately vague term; does it refer to *all* climate changes, natural and man-made, major and minor? This is never made clear.)

4. Rituals without Understanding

The New York Times writer John Tierney argues that the recycling of most types of materials is ultimately wasteful, given the enormous amount of energy and resources used in the recycling process, but the act of recycling "makes people feel virtuous, especially affluent people who feel guilty about their enormous environmental footprint. It is less an ethical activity than a religious ritual, like the ones performed by Catholics to obtain indulgences for their sins."[245]

The supposedly unconditional virtue of recycling has been "indoctrinated in students from kindergarten through college," and as a result, Tierney argues, "otherwise well-informed and educated people have no idea of the relative costs and benefits." The transportation and processing of consumer recyclables itself consumes a great deal of energy, and much of the resulting material is sent to landfills anyway.[246]

Because of the poor economics of recycling, hundreds of municipalities have stopped their recycling programs entirely, and this is not necessarily a bad thing. An economist has calculated that the total landfill size needed to contain all of the country's garbage for the next one thousand years amounts to less than 0.1% of the land available for grazing in the continental US.[247] Even that land would not be lost, since landfills have linings and other environmental safeguards, and once filled, they are typically covered with grass and converted to parkland.

Two other ritual-like behaviors that are encouraged today are: (1) turning out lights for "earth hour" (which has virtually no effect on climate), and (2) switching to reusable grocery bags (which raises hygiene concerns).

Given the above four Christianity-like elements within environmentalism, it should not be surprising that numerous Christians have made common cause with the movement. In 1979 Pope John Paul II sanctified Francis of Assisi as the patron saint of ecology for the Catholic Church. Christian theologians have started talking about "stewardship," while Al Gore talks about Creation Care and quotes from the Bible. Congregations are building "green churches" and celebrating Earth Day. The essential element in common between these two movements is the ethical ideal of altruism, which tells us to sacrifice ourselves for others. Greenism has simply expanded the definition of "others" to include the rest of the natural world.

Carbon Calvinism

PUBLISHER AND AUTHOR Peter Schwartz labels environmentalism as "the philosophy of privation," because it tells us that we must give up our comforts and the expectation of material progress.[248] If untouched earth is the standard of value, then, as human beings, virtually everything we do is bad. This ideology tells us to turn away from modern science and technology because they use up natural resources and distance us from nature. If we take this idea seriously, this leads to a shorter and vastly impoverished life: a life of hard work and an early death for most of us.

Environmentalism has always been more than just a social movement; it has always been a political movement as well. But a government that takes on the role of protecting the planet from humans is one that can control *everything* we do. It can control how much electricity we use, how much trash we dispose of, and how much we travel. It can control how much healthcare we use and how many children we have. In short, it can stop us from living. Writer Marion Tupy asks:

Is it really so difficult to imagine a future in which each of us is issued with a carbon credit at the start of each year, limiting what kind of food we eat (locally grown potatoes will be fine, but Alaskan salmon will be verboten) and how far we can travel (visiting our in-laws in Ohio once a year will be permitted, but not Paris)? In fact, it is almost impossible to imagine a single aspect of human existence that would be free from government interference—all in the name of saving the environment.[249]

Countless pseudo-scientific studies are done to see the effects of different actions on our "carbon footprints." An article in *New Scientist* suggests that the main problem with obesity is its carbon footprint; similar articles talk about the carbon cost of divorce and other social trends. The Swedish Ministry of Sustainable Development states that "women cause considerably fewer carbon dioxide emissions than men and thus considerably less climate change."[250]

This totalitarian aspect of the environmental movement has always appealed to former communists who were disenchanted by the clear failures of communism, and were looking for another ideology with which to oppose capitalism.

Science verses Quasi-Religion

As OPPOSED TO religions that rely on sacred texts, the sciences of ecology and climatology are primarily focused on reasoning from facts, as a perusal of their university textbooks should make clear.[251] Measurements are clearly showing an increasing global average temperature, and it is possible that most of the recent warming is due to increased emissions of carbon dioxide and other greenhouse gasses. Moreover, there is a potential for positive feedbacks to amplify a small warming into a much larger one, causing a significant sea-level rise and numerous other serious problems.

As we have seen in this chapter, there is also a decades-old, anti-humanist movement that is eerily similar to a religion (even though it does not worship a supernatural God), which is closely connected to these sciences. Many of the scientists have bought into this ideology.

The question then arises: To what extent should we hold these sciences in esteem, and respect their conclusions, and to what extent might they be corrupted by a religious and anti-humanist agenda? (It would certainly not be the first time that an entire scientific field was corrupted by ideology, as the history of Lysenkoism in the U.S.S.R. makes clear.)

Given the immense statistical problems and complexities of measuring climate phenomena, including the many types of poorly-understood positive and negative feedbacks, there are countless "judgment calls" that scientists have to make (such as setting criteria for rejection of temperature station readings due to the heat island effect). It is clearly possible for biases to influence reported scientific outcomes.

For an outsider to the environmental sciences, it is virtually impossible to know the extent of any corruption. However, there are at least three reasons for suspecting that this field generates biased conclusions that should not simply be taken at face value:

1. The term "climate change denier" is widely used to shut down discussion.
2. The "ClimateGate" scandal revealed corruption on the part of climate scientists.[252]
3. The IPCC global climate models' predictions have a poor track record.[253]

What should we do about this? We need to recognize, first, that the keys to human flourishing are reason, science, technology, and a political freedom that allows people to think and act on their ideas in pursuit of their own well-being. We also need to recognize that economic development and industry themselves (with the reliable energy sources on which they depend) provide us an enormous amount of protection from the vagaries of the natural climate (as well as any other conceivable existential threats to mankind, such as immense meteors and antibiotic-resistant superbugs).

We also must demand full accountability from climate scientists who predict doom and proscribe drastic political action. Drastic action requires dramatic justification. We need climate modelers to clearly and openly explain what their models predict, how reliable they have been, and how accurate we can expect them to be for predicting the future. We need climate scientists to openly address the concerns of their critics without dismissing them with ad-hominem arguments.

If it is clearly established that increasing carbon dioxide levels will cause terrible problems in the future (such as a nine-foot sea-level rise by 2100), then governments and private organizations may need to build carbon-capture facilities, and to transition from fossil fuels to low-carbon, reliable, high-energy-density sources such as nuclear and hydroelectric.[254] They may also need to build extra infrastructure and relocate the most impacted communities.

The title of this book is *God Versus Nature*, but given the meaning of "nature" accepted by the radical greens, the appropriate phrase is *Nature as God*.

CHAPTER FIFTEEN
Conclusion

H OW DOES THE history presented in this book relate to the philosophic issues of Chapter Four? That chapter identified three philosophic axioms: existence, consciousness, and identity. These can be stated as propositions:

- Existence exists. (Something *exists.*)
- Consciousness is conscious. (There is *awareness* of something which exists.)
- Existence is identity. (*To be* is to be *something.*)

The deepest philosophic root of the conflict between science and religion is metaphysical. It lies in two opposing views of the relationship between consciousness and existence: the *primacy of existence* versus the *primacy of consciousness.*

The *primacy of consciousness,* in its religious form, holds that a divine mind is the ultimate reality and source of explanation. This is the view that "God created the universe." Given this view, the question "what then created God?" is not considered a valid question; God is the starting point.

The *primacy of existence* holds that existence—a causal reality—is the ultimate reality and source of explanation. Given this view, "Who or what created existence?" is not a valid question, since existence is the starting point. It is improper to ask for the cause of existence as such; causes are *part* of existence. Our primary contact with existence is provided by our senses, which form the base on which we use logic and concepts to understand the world. These elements—the senses, concepts, and logic—are the key components of reason.

As we have seen, these philosophic ideas have dueled throughout the history of western civilization.

Science was born in the first predominantly secular society. Ancient Greece was the first culture in which the *primacy of existence* was largely accepted, especially in Aristotle's philosophy. Science ground to a halt in the highly religious Christian society of the Dark Ages, in which the *primacy of consciousness* held sway in intellectual life.

Science made a partial comeback in the fundamentally mixed society of Aristotle and Allah, and then science evaporated when this society re-asserted its Islamic roots, with an Ash'arite theology explicitly committed to the *primacy of consciousness.*

Aristotle's philosophy was rediscovered and became popular in late medieval Europe in large part thanks to Thomas Aquinas, and this led to the Renaissance. European Christian culture then acted as a carrier for the Aristotelian view of the significance of this world. Late medieval Christian culture contained two contradictory elements: the *primacy of existence* and the *primacy of consciousness.*

The Renaissance set the ground for such scientific thinkers as Copernicus, Kepler, Bacon, and Galileo—the symbol of truth standing up against authority. As we saw in Chapter Seven, the infamous clash between Galileo and the Church was indeed a clash between science and religion, in spite of historians' claims to the contrary.

During the Scientific Revolution, the new thinkers mentally changed gears and reduced the role of God in the universe, even as they re-asserted their religiosity. Newton identified the fact that "hypotheses" not grounded in evidence are not worth even considering, thereby unintentionally undercutting the basis of religion.

The new science of geology then turned against the Biblical account of the creation of the world in Genesis. This seriously undermined the Bible—the supposed voice of God's consciousness.

In the nineteenth century, Darwin discovered that *natural selection* was the basis of evolution. This was a powerful refutation of the popular field of natural theology, and it led to the re-evaluation of man as a purely natural being.

In Chapter Twelve, we saw that Kant's philosophy drove a wedge between reason and reality, enabling religionists to follow their faith with impunity, defending themselves with the claim "reason doesn't apply here."

In Chapter Thirteen, we saw how living life with morality and purpose is entirely compatible with a secular philosophy of reason.

Finally, in Chapter Fourteen, we saw how valuable modern industry is to our current and future well-being. Unfortunately, the modern environmental movement—which should be focused on the *human* environment—is infected with an anti-humanist quasi-religious ideology.

Religion and Modern Physics

IN THE TWENTIETH century, several developments in modern physics and cosmology have inspired some religionists to claim scientific support for religion. One of these is the discovery of evidence for a Big Bang that occurred ten to fifteen billion years ago, creating an expanding universe continuing until today. What caused the Big Bang? That is a question that we may never answer. But the claim that a consciousness is the cause is entirely arbitrary, and leads to the follow-up question, "What created this consciousness?"

Another development from physics has led to the "finely-tuned universe" being offered as a sign of God's design. A number of nonreligious physicists, in order to explain this so-called fine-tuning, have unfortunately felt compelled to posit an infinite number of universes, with each having different values of the fundamental physical constants. (We are fortunate enough to inhabit a universe with the right constants.) But there is no actual evidence for any universe other than this one; the idea of many different worlds is entirely arbitrary.

Why do these constants have the values they hold and not different values? The intellectually honest answer for the scientist is simply "We don't know." Perhaps we will know someday, as science has explained so many previously incomprehensible phenomena. Maybe we will never know. But we must bear in mind that *ignorance about the natural is never evidence for the supernatural.*

Science and Religion in the Twenty-First Century

SINCE THE NINETEENTH century, there has been a widespread de-coupling of science and religion, but there are still significant areas in which clashes continue. In school systems throughout America and Europe, the war over Darwin rages on, with increasing numbers of schools deciding to completely omit the topic of evolution (the core of modern biology) in order to avoid controversy.

In medicine, the field of medical ethics has been heavily influenced by religious ideas, affecting such issues as abortion, the use of stem cells, and end-of-life decisions.[255]

Influential leaders of the environmentalist movement continue to predict imminent climate catastrophe and damn human civilization as a plague, instead of promoting honest and open discussion of how mankind can best flourish now and into the future.

As long as people have commitments to both religion and to the comforts of modern technology, they will try to reconcile science and religion, no matter how contradictory the attempt.

As Douglas Murray points out in his book, *The Strange Death of Europe*, today's dominant secularists have no inspiring values to offer. A recent atheist bus slogan in Britain tells us, "There's probably no God. Now stop worrying and enjoy your life." Murray responds: "The question of how we are to enjoy that life is answered only with, 'However you see fit.' Who knows what will step into this void."[256]

Atheism itself is merely the rejection of God; it does not indicate any positives. What we need is a full-scale movement in support of a philosophy of reason. We need those who are committed to reason to confidently promote their ideas and inspire people to live their lives to the fullest—to achieve *eudaimonia*. Until such a movement occurs, we can expect religion and other irrational movements to remain dominant forces in shaping the future.

APPENDIX

John W. Draper, Andrew D. White, and the "Conflict Thesis"

$\mathbf{A}$s we saw in the book's introduction, academic historians of science coined the term "conflict thesis" for the idea that religion and science are fundamentally in conflict. This idea was popularized by two Americans living in the late nineteenth century, and it was subsequently attacked by academics in the late twentieth century.

The Identification of the Conflict

CHAPTER TEN EXPLAINED that Enlightenment thinkers such as the Marquis de Condorcet had seen that modern science relied on a causal view of nature, and that the existence of miracles was a direct contradiction to this view. This confidence in science led them to have scorn for the miraculous Biblical stories and to see an important conflict between science and religion:

> [Christianity] feared that spirit of doubt and inquiry, that confidence in one's own reason, which is the bane of all religious beliefs. The natural sciences were odious and suspect, for they are very dangerous to the success of miracles; and there is no religion that does not force its devotees to swallow a few physical absurdities.[257]

This was the conflict thesis, in embryo form. The idea was not explicitly articulated or systematically developed, and most scientists continued to believe that science and religion could be reconciled. It would take more scientific advances (in geology and biology) before the conflict thesis could be fully born and see the light of day. Moreover, these scientific advances would need to be noted by thinkers of an Enlightenment persuasion.

As described in Chapters Ten and Eleven, discoveries in the fields of geology and biology further eroded religious belief. Especially important was Darwin's theory of evolution by natural selection, which occasionally led to heated public conflicts.

Among the early public clashes between theologians and supporters of Darwin, one of the most publicized occurred in 1860 at a meeting of the British Association in Oxford. After the keynote address of the meeting discussed Darwin's theory of evolution, there was a confrontation between Bishop Samuel Wilberforce and Thomas H. Huxley (also known as "Darwin's bulldog"). According to reports, Wilberforce dismissed the idea of evolution and then asked Huxley if it was through his grandfather or his grandmother that he claimed his descent from a monkey. Huxley replied that he was not ashamed to have a monkey for his ancestor, but he would be ashamed to be connected with a man who used his eloquence to obscure the truth. One of those who witnessed this exchange was the man who had delivered the keynote address itself: an American scientist and educator named John William Draper.

Draper would later write an enormously popular book titled *History of the Conflict between Religion and Science,* published in 1874. Draper's theme was that "the history of science is not a mere record of isolated discoveries; it is a narrative of the conflict of two contending powers: the expansive force of human intellect and the compression arising from traditionary faith."[258]

After growing up in a Methodist household in England, Draper had been exposed in his college education to Auguste Comte's philosophy of Positivism. Impressed by Comte's high regard for science and denigration of religion, he turned away from his religious background and embraced an Enlightenment-style deism. After moving to America, he studied and lectured on both medicine and chemistry, and he became the first president of the American Chemical Society. In later years he turned to the subject of history, writing books on European history, the American Civil War, and the conflict between science and religion.

Draper believed the fundamental problem with religious dogma (especially Roman Catholic dogma) was its being essentially static and frozen, and incapable of refinement in the light of new knowledge: "A divine revelation must necessarily be intolerant of contradiction; it must repudiate all improvement in itself, and view with disdain that arising from the progressive intellectual development of man. But our opinions on every subject are continually liable to modification, from the irresistible advance of human knowledge."[259]

Draper's book traced science and Christianity from their origins, and found numerous examples of clashes in such subjects as cosmology, geology, anthropology, and history:

> The Church declared that the earth is the central and most important body in the universe; that the sun and moon and stars are tributary to it. On these points she was worsted by astronomy. She affirmed that a universal deluge had covered the earth; that the only surviving animals were such as had been saved in an ark. In this her error was established by geology. She taught that there was a first man, who, some six or eight thousand years ago, was suddenly created or called into existence in a condition of physical and moral perfection, and from that condition he fell. But anthropology has shown that human beings existed far back in geological time, and in a savage state but little better than that of the brute.[260]

Draper did claim allegiance to religion, but to him this meant a highly liberalized version of Christianity. He rejected the divine inspiration of the Bible and vigorously attacked Roman Catholicism, but he saw Protestantism as valuable when purged of its dogmatic "Catholic" components, and he accepted the Christian view of morality. In spite of Draper's claims of religiosity, his book was widely condemned as a vicious attack on religion as such. Perhaps because of the resulting notoriety, the book enjoyed great success in publication. By the 1930s, it had gone through fifty printings and had been translated into French, German, Italian, Dutch, Spanish, Polish, Japanese, Russian, Portuguese, and Serbian.

Draper's book was not the only of its kind. In 1896 it was followed by a similar work by another American: Andrew Dickson White. White had grown up in an Episcopalian household in upstate New York. Like Draper, he had been deeply influenced by Enlightenment ideals, and he later adopted the same type of deist religion. White saw religion (when appropriately purified) as the recognition of "a Power in the universe, not ourselves, which makes for righteousness." To be religious, for White, was to love God and one's neighbors, and little else was required. His career was divided between the realms of university education and government. As an educator, White was a professor of history and later became the first president of Cornell University, after working with Ezra Cornell to draw the charter of the new institution.

As one of the first completely secular universities, Cornell had no ties to any religious organizations. White had specified that "persons of every religious denomination, or of no religious denomination," were to be "equally eligible to all offices and appointments."[261] Because of this policy of secularism, the university was denounced as "godless" and White was attacked as an atheist and infidel.

In response, White began creating articles and lectures about the dangers that result when science and religion interfere with one another, and these gradually evolved into his major work: *A History of the Warfare of Science with Theology in Christendom*. White's book, in two volumes, was almost 900 pages long. Its twenty chapters addressed a wide range of subjects, including cosmology, geography, comets, geology, evolution, anthropology, meteorology, medicine, philology, political economy, and biblical criticism. In every subject, when science had been interfered with by "dogmatic theology," mankind was worse off because of it:

> In all modern history, interference with science in the supposed
> interest of religion, no matter how conscientious such interference
> may have been, has resulted in the direst evils both to religion and

to science, and invariably; and, on the other hand, all untrammeled scientific investigation, no matter how dangerous to religion some of its stages may have seemed for the time to be, has invariably resulted in the highest good both of religion and science.[262]

White's book never became a bestseller, but for a work of its size and scholarly content, it reached a wide audience. As one historian summarizes:

> No work — not even John William Draper's best-selling *History of the Conflict between Religion and Science* (1874) — has done more than White's to install in the public mind a sense of the adversarial relationship between science and religion. His *Warfare* remains in print to the present day, having appeared also in German, French, Italian, Swedish, and Japanese translations. His military rhetoric has captured the imagination of generations of readers.[263]

Draper and White were, without a doubt, the most influential champions of the "conflict thesis" in history, and their names continue to be mentioned whenever this conflict is discussed. Ironically, both of them claimed to be religious and denied that they were attacking religion as such. But, in essence, they *were* attacking religion. Both of them rejected faith as a means to knowledge and moral guidance, both saw science as an unqualified good, and both wrote books itemizing the countless clashes between religion and science throughout history.

The arguments of Draper and White, however, were weakened by their inconsistencies. To build a solid case for the conflict thesis, they would have needed the courage to disavow religion entirely. And they would have needed to identify two fundamental philosophic issues explicitly:

1. Reason—the faculty based on observation and logic, and the base of science—is the reality-oriented faculty, and cannot lead to belief in any supernatural entity such as God.
2. Faith—the reliance on the emotions as tools of cognition—is always invalid and subversive of reason.

The "conflict thesis" continued to be promoted well into the twentieth century, with titles such as James Y. Simpson's *Landmarks in the Struggle between Science and Religion* (1925) and Bertrand Russell's *Religion and Science* (1935). From the opposite perspective, a new American cultural movement ended up further popularizing the conflict thesis: A series of 12 tracts, called *The Fundamentals*, was published between 1910 and 1915, edited by a Baptist leader named A. C. Dixon. These inspired the movement known as evangelical fundamentalism, which was in part a reaction against social changes, including industrialization and the shock of World War I. The most significant factor in its rise was the growth of public education between 1900 and 1920. During this time, government-controlled secondary-level education quickly became widespread. In areas that had previously been extremely isolated, large numbers of people were suddenly exposed to modern science for the first time. The shock from the new ideas created the fundamentalist reaction. Laws were enacted to prevent the teaching of evolution. Publicity from the Scopes "Monkey" Trial of 1925 contributed to the awareness among Americans of fundamentalism and its conflict with science.

To this day, with stem-cell research and Intelligent Design in the news, clashes between science and religion are still widespread and obvious. It is no surprise that the conflict thesis is alive and well among non-intellectuals today. But this is not the case within the discipline of the History of Science.

The Fall of the Conflict Thesis

THE HISTORY OF Science as an academic discipline began early in the twentieth century, with its primary focus on understanding sequences of great scientific discoveries through history, and how these discoveries ended up changing our view of ourselves and of the universe in which we live. Also notable were biographies of the great scientists—the heroes who could inspire us with their achievements.

Early historians of science were impressed by the details and the comprehensiveness of White's book. However, as the works of Draper and White came under closer scrutiny, it became apparent that both contained numerous historical errors. These inaccuracies would later become focal points and excuses for historians of science to dismiss the accounts altogether.

One of these inaccuracies concerned the shape of the earth and the voyages of Columbus. Both Draper and White promoted the idea that before Columbus the earth was commonly believed to be flat, and that Columbus had proved them wrong. In fact, among educated Europeans, the idea of a spherical earth—promoted by the Ancient Greeks—had never been lost. The official Church cosmology followed the writings of Ptolemy, in which the earth was definitely spherical. There had been a few "flat-earthers," but these men had no following and were rejected by the Church.[264]

Another historical inaccuracy concerned the legality of human dissections during the Middle Ages. White promoted the idea that Church authorities had uniformly and explicitly forbidden human dissections throughout this period. Human dissections had been formally prohibited under non-Christian Roman rule, and in the following centuries they were not socially acceptable. But there were no widespread Church bans on such practices, and they started becoming common in universities during the thirteenth century.

The Role of Philosophy

THE HISTORICAL ERRORS of Draper and White were not enough by themselves to discredit the conflict thesis in the eyes of the historians. Something much more historically significant was at work. This was *philosophy*—specifically, the skepticism of modern post-Kantian philosophy.

The medieval philosophers, whatever their flaws, had clearly understood the difference between reason and faith. Reason could not answer all questions, they held, which is why they needed faith. But for those questions that reason could answer, it did yield knowledge. The medieval philosophers were not skeptics.

Immanuel Kant, to the contrary, introduced the idea that man's reason is a delusion. Our reason, claimed Kant, is structured in such a way that we all necessarily see the world through certain "categories," which do not yield knowledge of reality "as it really is." That is, we can know nothing, period.

"Faith" is considered the method we need for things we cannot know through reason. But if we can know nothing through reason to begin with, there is no basis for the distinction. Put another way, whatever a person claims as "knowledge" through a delusional mental mechanism, his knowledge has no greater validity than the dogmas of faith.

Most historians (like most people) do not make such premises of their philosophy clear to themselves. The premises operate nonetheless when they practice their craft. When a historian influenced by Kantian-based philosophy confronts an issue of reason and faith in history, he is bound to blur the difference between the two; the distinction between reason and faith, to him, is virtually meaningless.

* * *

HISTORY, AS A discipline, relies on basic premises about the nature of reality, knowledge, and values—i.e., it relies on philosophy. But philosophers have been relentlessly attacking reason—which is the base of science. When philosophers can no longer tell the difference between reason and faith, it is not surprising that our historians cannot either, and that they cannot see any conflict between science and religion.

The conflict thesis was grounded in the Enlightenment's confidence in reason and distrust of religion. While in Europe the Enlightenment faded quickly into the irrationalism of the romanticists, in America the Enlightenment ideals had a greater momentum, and so it was two nineteenth-century Americans—John William Draper and Andrew Dickson White—who became the most influential advocates of the conflict thesis. Since this thesis never had a proper philosophic defense, it could not survive the twentieth-century assault on reason.

The relationship between science and religion, as seen through history, is undoubtedly complex. However, this complexity can only be managed by means of principles. A proper understanding of the relationship between science and religion must rest on philosophic fundamentals, i.e., a proper understanding of reason as the reality-oriented faculty, faith as the reliance on emotions as tools of cognition, and the consequent incompatibility of reason and faith. Only when the issue is seen in these terms will modern historians of science start genuinely contributing to our knowledge of this fascinating subject.

ACKNOWLEDGMENTS

C HAPTER EIGHT IS an edited version of my article "The Role of Religion in the Scientific Revolution," published in The Objective Standard, Fall 2012.

During the process of writing this book, I have received valuable assistance from many individuals, to whom I am grateful. Eric Allison provided extensive comments on an early draft. Steve Jolivette provided comprehensive feedback over multiple iterations of various chapters. Numerous members of a Northern Virginia book discussion group also provided valuable feedback. Bruce Van Horne did proofreading of the final copy.

And to my wife Roksana—the love of my life—thank you for having done so much!

SELECT BIBLIOGRAPHY

Aristotle. *The Nicomachean Ethics, The Basic Works of Aristotle.* Edited by Richard McKeon. Translated by W. D. Ross. New York: Random House, 1941.

Baxter, Stephen. *Revolutions in the Earth: James Hutton and the True Age of the World.* London: Phoenix, 2003.

Binswanger, Harry. *How We Know: Epistemology on an Objectivist Foundation.* New York: TOF Publications, 2014.

Bowler, Peter J. *Evolution: The History of an Idea.* 3rd ed. Berkeley: University of California Press, 2003.

Clagett, Marshall. *Greek Science in Antiquity.* New York: Abelard-Schuman, 1955.

Epstein, Alex. *The Moral Case for Fossil Fuels.* New York: Penguin, 2014.

Freeman, Charles. *The Closing of the Western Mind: The Rise of Faith and the Fall of Reason.* New York: Alfred A. Knopf, 2003.

Grant, Edward. *The Foundations of Modern Science in the Middle Ages: Their Religious, Institutional, and Intellectual Contexts.* Cambridge: Cambridge University Press, 1996.

Gingras, Yves. *Science and Religion: An Impossible Dialogue.* Translated by Peter Keating. Cambridge: Polity Press, 2017.

Gotthelf, Allan, and Gregory Salmieri, eds. *A Companion to Ayn Rand.* West Sussex: John Wiley & Sons, 2016.

Harriman, David. *The Logical Leap: Induction in Physics*. New York: New American Library, 2010.

Herman, Arthur. *The Cave and the Light: Plato Versus Aristotle, and the Struggle for the Soul of Western Civilization*. New York: Random House, 2013.

Jones, W. T. *A History of Western Philosophy*, 2nd ed. New York: Harcourt Brace Jovanovich, 1969.

Lindberg, David C. *The Beginnings of Western Science: The European Scientific Tradition in Philosophical, Religious, and Institutional Context, Prehistory to A.D. 1450*. 2nd ed. Chicago: University of Chicago Press, 2007.

Numbers, Ronald L. *The Creationists: From Scientific Creationism to Intelligent Design*. Expanded ed. Cambridge: Harvard University Press, 2006.

Peikoff, Leonard. "The History of Philosophy, Volume 1 – Founders of Western Philosophy: Thales to Hume." Lecture course. Recorded in 1972. Ayn Rand Institute eStore.

———. "The History of Philosophy, Volume 2 – Modern Philosophy: Kant to the Present." Lecture course. Recorded in 1970. Ayn Rand Institute eStore.

———. *Objectivism: The Philosophy of Ayn Rand*. New York: Meridian, 1991.

Pinker, Steven. *Enlightenment Now: The Case for Reason, Science, Humanism, and Progress*. New York: Viking, 2018.

Rand, Ayn. *For the New Intellectual: The Philosophy of Ayn Rand*. New York: Penguin, 1963.

———. *Introduction to Objectivist Epistemology*. Expanded 2nd ed. Edited by Harry Binswanger and Leonard Peikoff. New York: Meridian, 1990.

———. *The Return of the Primitive: The Anti-Industrial Revolution*. New York: Meridian, 1999.

————. *The Virtue of Selfishness: A New Concept of Egoism.* New York: Signet, 1964.

Reilly, Robert R. *The Closing of the Muslim Mind: How Intellectual Suicide Created the Modern Islamist Crisis.* Wilmington: ISI Books, 2010.

Rheins, Jason. *The Philosophy of Immanuel Kant, Part I: Theoretical Philosophy.* Lecture. Recorded in 2007 at the Objectivist Summer Conference. Ayn Rand Institute eStore.

Rubenstein, Richard E. *Aristotle's Children: How Christians, Muslims, and Jews Rediscovered Ancient Wisdom and Illuminated the Dark Ages.* New York: Harcourt, 2003.

Sobel, Dava. *Galileo's Daughter: A Historical Memoir of Science, Faith, and Love.* New York: Walker, 1999.

Smith, Tara, *Ayn Rand's Normative Ethics: The Virtuous Egoist.* Cambridge: Cambridge University Press, 2006.

————. *Viable Values: A Study of Life as the Root and Reward of Morality.* New York: Rowman & Littlefield: 2000.

Stove, David. *Scientific Irrationalism: Origins of a Postmodern Cult.* London: Transaction, 2007.

Westfall, Richard S. *Science and Religion in Seventeenth-Century England.* Ann Arbor: University of Michigan Press, 1973.

Zubrin, Robert. *Merchants of Despair: Radical Environmentalists, Criminal Pseudo-Scientists, and the Fatal Cult of Antihumanism.* New York: Encounter Books, 2012.

NOTES

[1] Peter J. Bowler and Iwan Rhys Morus, *Making Modern Science* (Chicago: University of Chicago Press, 2005), 364.

[2] Richard G. Olson, *Science and Religion: From Copernicus to Darwin* (Baltimore: Johns Hopkins University Press, 2004), 221.

[3] Colin A. Russell, "The Conflict of Science and Religion," in *Science and Religion*, ed. Gary B. Ferngren (Baltimore: Johns Hopkins University Press, 2002), 7–8.

[4] Lawrence M. Principe, transcript of lecture course "Science and Religion" (Chantilly, VA: The Teaching Company, 2006), 23.

[5] Ruth Harris, *Lourdes: Body and Spirit in the Secular Age* (New York: Viking, 1999), 3–4.

[6] Ibid., 9–10.

[7] Andrew Gregory, *Harvey's Heart: The Discovery of Blood Circulation* (Cambridge: Icon Books, 2001), 60–1.

[8] Albert S. Lyons and R. Joseph Petrucelli II, *Medicine: An Illustrated History* (New York: Abrams, 1987), 434.

[9] This statement is based on a definition offered by John Ridpath in his excellent lecture set "Religion vs. Man," recorded in 1989 at The Jefferson School Conference.

[10] Leonard Peikoff, "Religion vs. America," in *The Voice of Reason: Essays in Objectivist Thought*, ed. Leonard Peikoff (New York: Penguin, 1989), 67–8.

[11] Tim M. Berra, *Evolution and the Myth of Creationism: A Basic Guide to the Facts in the Evolution Debate* (Stanford: Stanford University Press, 1990), 2.

[12] David Harriman, *The Logical Leap: Induction in Physics* (New York: New American Library, 2010), 24.

[13] Peter Whitfield, *Landmarks in Western Science: From Prehistory to the Atomic Age* (New York: Routledge, 1999), 22.

[14] Ibid., 26.

[15] Thomas Cahill, *Sailing the Wine-Dark Sea: Why the Greeks Matter* (New York: Anchor, 2004), 235.

[16] Leonard Peikoff, "Religion vs. America," in *The Voice of Reason: Essays in Objectivist Thought* by Ayn Rand (New York: Penguin, 1989), 69.

[17] Whitfield, *Landmarks in Western Science*, 28.

[18] G. E. R. Lloyd, *Early Greek Science: Thales to Aristotle* (New York: Norton, 1970), 8.

[19] Quoted in Edward Grant, *A History of Natural Philosophy: From the Ancient World to the Nineteenth Century* (Cambridge: Cambridge University Press, 2007), 21.

[20] Quoted in Lloyd, *Early Greek Science*, 54.

[21] Cf. Leonard Peikoff, "Epilogue: The Duel Between Plato and Aristotle," in *Objectivism: The Philosophy of Ayn Rand* (New York: Dutton, 1991), 451–8.

[22] For dates, I have decided to use the B.C./A.D. notation, instead of the currently popular B.C.E./C.E. "common era" notation. As Scott Powell argues, "We must dispense with this revisionist usage. . . . There is no objective justification for it. Human civilization experienced no union in any regard 1492 years or so before Columbus to generate a 'common' story." See Scott Powell, *The History of Now: A New Kind of History* (Scott Powell, 2018), 22.

[23] Plato, *Republic*, trans. G. M. A. Grube, rev. C. D. C. Reeve (Indianapolis: Hackett, 1992), 186–8 (Bk. VII, lines 514–6).

[24] Leonard Peikoff, *The DIM Hypothesis: Why the Lights of the West Are Going Out* (New York: New American Library, 2012), 23–4.

[25] Plato, *Five Dialogues: Euthyphro, Apology, Crito, Meno, Phaedo*, trans. G. M. A. Grube (Indianapolis: Hacket, 1981), 103 (*Phaedo* 66e).

[26] Plato, *Republic*, 202.

[27] Peikoff, *DIM Hypothesis*, 28.

[28] Quoted in Grant, *History of Natural Philosophy*, 44.

[29] Peikoff, *DIM Hypothesis*, 30.

[30] Jonathan Barnes, *Aristotle: A Very Short Introduction* (Oxford: Oxford University Press, 2000), 18.

[31] Ibid., 137.

[32] Peikoff, *DIM Hypothesis*, 194–5.

[33] See, for example, the discussion of Demetrius, the onetime ruler of Athens and student of Aristotle who eventually settled in Alexandria in Luciano Canfora, *The Vanished Library: A Wonder of the Ancient World* (Berkeley: University of California Press, 1990).

[34] David C. Lindberg, *The Beginnings of Western Science: The European Scientific Tradition in Philosophical, Religious, and Institutional Context, Prehistory to A.D. 1450*, 2nd ed. (Chicago: University of Chicago Press, 2007), 129.

[35] Whitfield, *Landmarks in Western Science*, 36.

[36] Stephen Bertman, *The Genesis of Science: The Story of Greek Imagination* (Amherst, New York: 2010), 62.

[37] An excellent animation explaining Eudoxus's model can be found at youtube.com/watch?v=_SFzDYSqR_4

[38] Whitfield, *Landmarks in Western Science*, 38.

[39] The story of the history of the Antikythera mechanism is engagingly told in Jo Marchant's *Decoding the Heavens: A 2,000-Year-Old Computer—And the Century-Long Search to Discover its Secrets* (Cambridge, MA: Da Capo Press, 2009).

[40] Marchant, *Decoding the Heavens*, 259–60.

[41] A good figure showing the epicycles, the deferent, and the equant can be found at commons.wikimedia.org/wiki/Category:Ancient_Greek_Astronomy#/media/File:Ptolemaic_deferent_and_epicycle_taking_into_account_the_three_centres.svg

[42] "Ptolemy" entry, *Dictionary of Scientific Biography* (New York: Scribner, 1970), 196.

[43] Harriman, *Logical Leap*, 25.

[44] W. T. Jones, *A History of Western Philosophy: The Classical Mind*, 2nd ed. (New York: Harcourt Brace Jovanovich, 1970), 321.

[45] The many different arguments of the Skeptics are known to us primarily through the writings of Sextus Empiricus (c. A.D. 200), who collected and wrote down all the main Skeptic arguments; he is the main source on what's known of the Skeptic school.

[46] Transcribed by the author from lecture six of *The History of Philosophy: Founders of Western Philosophy: Thales to Hume*, by Leonard Peikoff (Irvine, CA: Ayn Rand Institute Bookstore, 1994).

[47] Edward Grant, *Science and Religion, 400 B.C. to A.D. 1550: From Aristotle to Copernicus* (Baltimore: Johns Hopkins University Press, 2004), 99–100. Here we see the roots of the Neoplatonic answer to the problem of evil. This problem is

framed as follows: If the ultimate source and controller of reality, the One—or God—is good, then why do bad things happen in the world? The Neoplatonic answer is that physical matter is the "darkest," "lowest" level of existence. It is unreal, absence, non-being. It is in effect where the emanating rays of the One run out. So one should not expect much "goodness" in this world.

[48] Jones, *History of Western Philosophy: The Medieval Mind*, 2nd edition (New York: Harcourt Brace Jovanovich, 1969), 2.

[49] Ibid., 31.

[50] 1 Cor. 3.19.

[51] Translation from Col. 2:8 by David C. Lindberg, "Myth 1: That the Rise of Christianity Was Responsible for the Demise of Ancient Science," in *Galileo Goes to Jail: and Other Myths about Science and Religion*, ed. Ronald L. Numbers (Cambridge: Harvard University Press, 2009), 10. This book contains many good examples of the kinds of academic argumentation that I criticize.

[52] Charles Freeman, *The Closing of the Western Mind: The Rise of Faith and the Fall of Reason* (New York: Alfred A. Knopf, 2003), 316.

[53] David C. Lindberg, "Science and the Early Church," in *God and Nature: Historical Essays on the Encounter between Christianity and Science*, ed. David C. Lindberg and Ronald L. Numbers (Berkeley: University of California Press, 1986), 25.

[54] Tertullian, *On the Flesh of Christ*, trans. Dr. Holmes, ebook version (OrthodoxEbooks), 943. Some historians have tried to defend this statement as an expression of Aristotelian rhetoric, but I cannot take this defense seriously. For example, David C. Lindberg argues that "Tertullian was simply making use of a standard Aristotelian argumentative form, maintaining that the more improbable an event, the less likely is anybody to believe, without compelling evidence, that it has occurred; therefore, the very improbability of an alleged event, such as Christ's resurrection, is evidence in its favor." (Lindberg, "Science and the Early Church", 26.)

[55] Saint Ambrose, *Hexameron, Paradise, and Cain and Abel*, trans. John J. Savage (New York: Fathers of the Church, 1961), 20–1.

[56] Ibid., 46.

[57] Ibid., 9.

[58] Augustine, *On the literal meaning of Genesis*, trans. John Hammond Taylor (New York: Paulist Press, 1982), 52 (Book 2, Chapter 5, final paragraph).

[59] Quoted in Lindberg, "Science and the Early Church," 31.

⁶⁰ David C. Lindberg, "The Medieval Church Encounters the Classical Tradition: Saint Augustine, Roger Bacon, and the Handmaiden Metaphor," in *When Science and Christianity Meet*, ed. David C. Lindberg and Ronald L. Numbers (Chicago: University of Chicago Press, 2003), 15.

⁶¹ Quoted in Lindberg, "Medieval Church," 14–15.

⁶² Augustine, *Confessions*, trans. R. S. Pine-Coffin (London: Penguin, 1961), 93.

⁶³ Ibid., 241.

⁶⁴ Quoted in Freeman, *Western Mind*, 193–4.

⁶⁵ Freeman, *Western Mind*, 214.

⁶⁶ Ibid., 268–9.

⁶⁷ William Manchester, *A World Lit Only By Fire: The Medieval Mind and the Renaissance: Portrait of an Age* (Boston: Little, Brown and Company, 1992), 3.

⁶⁸ Ayn Rand, *The Virtue of Selfishness: A New Concept of Egoism* (New York: Penguin, 1964), 22.

⁶⁹ A full defense of reason would need to address such topics as the validity of perception, the nature of concepts, definitions, propositions, objectivity, logic, and volition. For good discussions of these issues, see *Introduction to Objectivist Epistemology* by Ayn Rand, and *How We Know: Epistemology on an Objectivist Foundation*, by Harry Binswanger.

⁷⁰ Paul Feyerabend, *Against Method*, rev. ed. (London: Verso, 1988), 36.

⁷¹ Andrew Bernstein, *Objectivism in One Lesson: An Introduction to the Philosophy of Ayn Rand* (New York: Hamilton Books, 2008), 37–8.

⁷² Ayn Rand, *Atlas Shrugged* (Random House: New York, 1957), 1015.

⁷³ Consciousness is a biological faculty of certain living organisms. While it cannot directly affect reality outside an organism, it can cause the organism to take actions that affect the world. It is only in this sense that consciousness can change reality.

⁷⁴ Bernstein, *Objectivism in One Lesson*, 38.

⁷⁵ Ayn Rand, *Introduction to Objectivist Epistemology*, exp. 2nd ed. (New York: Meridian, 1990), 59.

⁷⁶ Rand, *Atlas Shrugged*, 1015.

⁷⁷ Leonard Peikoff, *Objectivism: The Philosophy of Ayn Rand* (New York: Penguin, 1991), 18. For a more detailed treatment of this idea, see Jason Rheins's chapter "Objectivist Metaphysics: The Primacy of Existence," in *A Companion to Ayn Rand*, ed. A. Gotthelf and G. Salmieri (West Sussex: John Wiley & Sons, 2016).

[78] For a more involved discussion on this point, see Peikoff, *Objectivism*, 7–12, 245–71.

[79] Leonard Peikoff, "Religion vs. America" in *The Voice of Reason: Essays in Objectivist Thought*, edited by Leonard Peikoff (New York: Penguin, 1989), 66.

[80] William Durant, *The Age of Faith: A History of Medieval Civilization—Christian, Islamic, and Judaic—from Constantine to Dante: A.D. 325–1300* (New York: Simon and Schuster, 1950), 249–50.

[81] Jim Al-Khalili, *The House of Wisdom: How Arabic Science Saved Ancient Knowledge and Gave Us the Renaissance* (New York: Penguin Books, 2010), 120.

[82] Quoted in Al-Khalili, *House of Wisdom*, 150.

[83] Quoted in David C. Lindberg, *The Beginnings of Western Science: The European Scientific Tradition in Philosophical, Religious, and Institutional Context, Prehistory to A.D. 1450*, 2nd ed. (Chicago: University of Chicago Press, 2007), 188.

[84] Toby E. Huff, *The Rise of Early Modern Science: Islam, China, and the West*, 2nd ed. (Cambridge: Cambridge University Press, 2003), 52.

[85] Robert R. Reilly, *The Closing of the Muslim Mind: How Intellectual Suicide Created the Modern Islamist Crisis* (Wilmington, DE: ISI Books, 2010), 22.

[86] Majid Fakhry, *A History of Islamic Philosophy*, 3rd ed. (New York: Columbia University Press, 2004), xix.

[87] Reilly, *Muslim Mind*, 51.

[88] Reilly, *Muslim Mind*, 43.

[89] Andrew Bernstein, "Great Islamic Thinkers Versus Islam," *The Objective Standard* 7, no. 4.

[90] Quoted in Reilly, *Muslim Mind*, 62–3.

[91] Toby E. Huff, *Intellectual Curiosity and the Scientific Revolution: A Global Perspective* (Cambridge: Cambridge University Press, 2011), 133.

[92] Ibid., 133–4.

[93] John Man, *Gutenberg: How One Man Remade the World with Words* (New York: MJF Books, 2002), 247–248.

[94] Francis Robinson, "Islam and the Impact of Print in South Asia," in *The Transmission of Knowledge in South Asia: Essays on Education, Religion, History, and Politics*, ed. Nigel Crook (Delhi: Oxford University Press, 1996), 65.

[95] Ibid., 65.

[96] Ibid., 69.

[97] Reilly, *Muslim Mind*, 161.

[98] Augustine, *Confessions*, trans. R. S. Pine-Coffin (London: Penguin, 1961), 169.

[99] Quoted in Richard E. Rubenstein, *Aristotle's Children: How Christians, Muslims, and Jews Rediscovered Ancient Wisdom and Illuminated the Dark Ages* (New York: Harcourt, 2003), 156.

[100] Rubenstein, *Aristotle's Children*, 8.

[101] This view emerges from Aristotle's *Physics*, especially books I & VIII.

[102] See Aristotle's *On the Soul*, in which he argues that the soul is essentially the form of its body.

[103] Aristotle, *The Basic Works of Aristotle*, ed. Richard McKeon (NY: Random House, 1941), 992 (*Nicomachean Ethics*, Bk. IV, Ch. 3).

[104] Quoted in Edward Grant, *A History of Natural Philosophy: From the Ancient World to the Nineteenth Century* (Cambridge: Cambridge University Press, 2007), 143.

[105] Lindberg, *The Beginnings of Western Science*, 227.

[106] Ibid., 223.

[107] Rubenstein, *Aristotle's Children*, 80.

[108] J. A. Weisheipl, "Albert the Great, St." entry, *New Catholic Encyclopedia*, 2nd ed. (New York: Gale, 2003), 226.

[109] Rubenstein, *Aristotle's Children*, 198.

[110] Grant, *Natural Philosophy*, 249.

[111] Charles B. Schmitt, *Aristotle and the Renaissance* (Cambridge: Harvard University Press, 1983), 14.

[112] *Martin Luther: Selections from his Writings*, ed. and with an Introduction by John Dillenberger (New York: Doubleday, 1962), 471.

[113] Schmitt, *Aristotle and the Renaissance*), 11.

[114] Paul Johnson, *The Renaissance: A Short History* (New York: Modern Library, 2000), 128.

[115] Ibid., 51.

[116] Grant, *Natural Philosophy*, 245.

[117] David Harriman, "Galileo: Inaugurating the Age of Reason," *The Intellectual Activist*, March 2000, 18.

[118] Quoted in Harriman, "Galileo," 21.

[119] James Reston, *Galileo: A Life* (Washington: Beard Books, 2000), 54.

[120] Pietro Redondi, *Galileo Heretic* (Princeton: Princeton University Press, 1987), 5–7.

[121] Harriman, "Galileo," 15.

[122] Maurice A. Finocchiaro, *The Galileo Affair: A Documentary History* (Berkeley: University of California Press, 1989), 292.

[123] Quoted in Giorgio de Santillana, *The Crime of Galileo* (Chicago: University of Chicago Press, 1955), 351.

[124] Reston, *Galileo*, 284.

[125] See the appendix for more on the problems of modern academic history of science.

[126] Maurice A. Finocchiaro, "Science, Religion, and the Historiography of the Galileo Affair," in *Osiris* 16 (2001), 128.

[127] Reston, *Galileo*, 267–8.

[128] I. Bernard Cohen, *Revolution in Science* (Cambridge: Harvard University Press, 1985), 79.

[129] Francis Bacon, *The Advancement of Learning*, ed. Stephen Jay Gould (New York: Modern Library, 2001), 10 (The First Book, I. 3).

[130] Francis Bacon, *Novum Organum*, trans. Peter Urbach and John Gibson (Chicago: Open Court, 1994), 72 (Book 1, Aphorism 65).

[131] Frank E. Manuel, *The Religion of Isaac Newton: The Freemantle Lectures* (Oxford: Clarendon Press, 1974), 30.

[132] Ibid., 30–1.

[133] Rose-Mary Sargent, *The Diffident Naturalist: Robert Boyle and the Philosophy of Experiment* (Chicago: University of Chicago Press, 1995), 114.

[134] Manuel, *Religion of Isaac Newton*, 30.

[135] Quoted in Richard S. Westfall, *Science and Religion in Seventeenth-Century England* (Ann Arbor, MI: University of Michigan Press, 1958), 31.

[136] Ibid., 27.

[137] Ibid., 125–6.

[138] Rodney Stark, *For the Glory of God: How Monotheism Led to Reformations, Science, Witch-hunts, and the End of Slavery* (Princeton: Princeton University Press, 2003), 147.

[139] From Newton's manuscript entitled "A Scheme for Establishing the Royal

Society." See *Newton's Philosophy of Nature: Selections from his Writings*, ed. H. S. Thayer (Mineola, NY: Dover, 2005), 1, 181

[140] Hereafter, I refer to this work as simply *Principia*.

[141] Isaac Newton, *The Principia: Mathematical Principles of Natural Philosophy, A New Translation* by I. Bernard Cohen and Anne Whitman (Berkeley, CA: University of California Press, 1999), 416–7.

[142] Newton, *Principia*, 943.

[143] I. Bernard Cohen points out that Newton was not saying that we should not try to find a deeper explanation for the law of gravity. See his discussion of this issue in Newton, *Principia*, 274–80.

[144] *Newton's Philosophy of Nature: Selections from His Writings*, ed. H. S. Thayer (New York: Hafner, 1953), 6.

[145] Newton, *Principia*, 943.

[146] Carl Sagan presents a good example of an arbitrary idea (although not using this term) in the chapter "The Dragon in my Garage," in his book *The Demon-Haunted World: Science as a Candle in the Dark* (New York: Ballantine Books, 1996), 171–3.

[147] For a good discussion of this law, see Lionel Ruby, *Logic: An Introduction* (Cresskill, NJ: The Paper Tiger, 2000), 131–45.

[148] Harry Binswanger, *How We Know: Epistemology on an Objectivist Foundation* (New York: TOF Publications, 2014), 178.

[149] A good explanation of this point is found in Peikoff, *Objectivism*, 163–171.

[150] Binswanger points out that there are two very different meanings of the word "possible," corresponding to epistemic possibility and metaphysical possibility. Here it should be clear that I am using the epistemic meaning. See Binswanger, *How We Know*, 276–7.

[151] In philosophy, the idea that "you are just a brain in a vat" is completely arbitrary. For a good example of the arbitrary in law, see Leonard Peikoff's 1996 lecture, "A Philosopher Looks at the O. J. Verdict."

[152] Ayn Rand, *For the New Intellectual* (New York: New American Library, 1961), 55.

[153] Ibid., 128.

[154] Antoine-Nicolas de Condorcet, *Sketch for a Historical Picture of The Progress of the Human Mind*, trans. June Barraclough (Westport, CT: Greenwood Press, 1955), 72.

[155] Gen. 1:1–5, King James Bible, Revised Standard Version.

[156] This belief came from ancient Jewish sources.

[157] Quoted in Claude C. Albritton, *The Abyss of Time: Changing Conceptions of the Earth's Antiquity After the Sixteenth Century* (New York: Dover, 1980), 38.

[158] Ibid., 59.

[159] Yves Gingras, *Science and Religion: An Impossible Dialogue*, trans. Peter Keating (Cambridge: Polity Press, 2017), 107–9

[160] For examples of igneous dikes, see wikipedia.org/wiki/Dike_(geology)

[161] For examples of unconformities, see wikipedia.org/wiki/Hutton%27s_Unconformity

[162] Quoted in Stephen Baxter, *Revolutions in the Earth: James Hutton and the True Age of the World* (London: Phoenix, 2004), 140.

[163] Ibid., 163.

[164] Quoted in Patrick Wise Jackson, *The Chronologers' Quest: The Search for the Age of the Earth* (Cambridge: Cambridge University Press, 2006), p.130.

[165] Quoted in James A. Secord's introduction to Charles Lyell, *Principles of Geology*, *edited and with an introduction by James A. Secord* (London: Penguin, 1997), xxiv.

[166] Quoted in Baxter, *Revolutions in the Earth*, 208.

[167] Charles Darwin, *On the Origin of Species, A Facsimile of the First Edition* (Cambridge: Harvard University Press, 1964), 282.

[168] Quoted in Keith Thomson, *Before Darwin: Reconciling God and Nature* (London: Yale University Press, 2005), 10–11.

[169] Edward J. Larson, *Evolution: The Remarkable History of a Scientific Theory* (New York: Modern Library, 2004), 40.

[170] Michael White and John Gribbin, *Darwin: A Life in Science* (New York: Penguin, 1995), 80.

[171] The fallacy of Malthusianism has been discussed in many places. For a recent critique, see Robert Zubrin, *Merchants of Despair: Radical Environmentalists, Criminal Pseudo-Scientists, and the Fatal Cult of Antihumanism* (New York: Encounter Books, 2012).

[172] White and Gribbin, *Darwin*, 200.

[173] Charles Darwin, *Autobiography*, ed. Nora Barlow (New York: Norton, 1958), 120.

[174] White and Gribbin, *Darwin*, 160.

[175] There has been a report that Darwin converted back to Christianity on his

deathbed, but I do not consider the original (and only) source of this report to be credible. See chapter 16 of *Galileo Goes to Jail: and Other Myths about Science and Religion*, ed. Ronald L. Numbers (Cambridge: Harvard University Press, 2009).

[176] Edward J. Larson, *Evolution: The Remarkable History of a Scientific Theory* (New York: Modern Library, 2004), 237.

[177] Ibid., 205.

[178] Ibid., 204.

[179] Ibid., 202.

[180] Ronald L. Numbers, *The Creationists: From Scientific Creationism to Intelligent Design*, exp. ed. (Cambridge: Harvard University Press, 2006), 97.

[181] Ibid., 215.

[182] Epperson v. Arkansas (1968) 393 U.S. 97, 37 U.S. Law Week 4017, 89 S. Ct. 266, 21 L. Ed 228.

[183] Quoted in Tim Berra, *Evolution and the Myth of Creationism: A Basic Guide to the Facts in the Evolution Debate* (Stanford: Stanford University Press, 1990), 5.

[184] Edwards v. Aguillard (1987) 482 U.S. 578.

[185] I owe the phrase "creationism in camouflage" to a lecture given in 2005 by Dr. Keith Lockitch of the Ayn Rand Institute, titled "Creationism in Camouflage: The 'Intelligent Design' Deception." I also owe numerous points to Dr. Lockitch's lecture.

[186] See *The Mystery of Life's Origins* (1984), written by three Protestant scientists, and *Evolution: Theory in Crisis* (1986) by the geneticist Michael Denton.

[187] Quoted in Numbers, *The Creationists*, 391–2.

[188] Percival Davis and Dean H. Kenyon, *Of Pandas and People: The Central Question of Biological Origins*, 2nd ed. (Dallas: Haughton, 1989), 14.

[189] Quoted in Numbers, *The Creationists*, 394.

[190] Using Hume's type of argument, we could ask how we know that our minds will continue to do this synthesizing activity in the same way tomorrow. Kant does not seem to have an answer to this.

[191] Stephen Hicks, *Explaining Postmodernism: Skepticism and Socialism from Rousseau to Foucault.* (Ockham's Razor, 2018), 28.

[192] Ibid., 29.

[193] Immanuel Kant, *Critique of Pure Reason*, 1781, trans. Norman Kemp Smith (New York: MacMillan, 1929), Bxxx.

[194] Kant's famous phrase is "God, freedom, and immortality." In spite of the inclusion of free will here, it is important to note that there have been entirely nonreligious defenses of free will. See Binswanger, *How We Know* and Edwin A. Locke, *The Illusion of Determinism: Why Free Will is Real and Causal* (Edwin A. Locke, 2017).

[195] Bo Dragsdahl, *Karl Popper's Assault on Science*, recorded lecture course (Irvine, CA: Ayn Rand Institute eStore, 2003).

[196] David Stove, *Scientific Irrationalism: Origins of a Postmodern Cult* (London: Transaction, 2007), 95.

[197] Paul Feyerabend, *Against Method*, rev. ed. (New York: Verso, 1988), 36.

[198] Thomas S. Kuhn, *The Structure of Scientific Revolutions* (Chicago: University of Chicago Press, 1970), 158.

[199] J. L. Heilbron, *Elements of Early Modern Physics* (Berkeley, CA: University of California Press, 1982), viii.

[200] Rand, *Objectivist Epistemology*, 81. For Rand's take on the significance of Kant's philosophy, see her article "From the Horse's Mouth," in *Philosophy: Who Needs It* (New York: Signet, 1984).

[201] Binswanger, *How We Know*, 94.

[202] Aristotle, *Nichomachean Ethics*, *The Basic Works of Aristotle*, ed. Richard McKeon, trans. W. D. Ross (New York: Random House, 1941), 1169a12.

[203] Rand, *Virtue of Selfishness*, 14.

[204] This section is my brief summary of a topic that has been covered in much more detail elsewhere. Rand's crucial (but condensed) essay is "The Objectivist Ethics," in her book *The Virtue of Selfishness: A New Concept of Egoism*. A fuller presentation can be found as Chapter 7, "The Good," in Leonard Peikoff's *Objectivism: The Philosophy of Ayn Rand*. For an accessible introduction to Rand's ethics, see Craig Biddle, *Loving Life: The Morality of Self-Interest and the Facts That Support It* (Glen Allen, VA: Glen Allen Press, 2002). For a comparison of Rand's metaethics to other theories, see Tara Smith, *Viable Values: A Study of Life as the Root and Reward of Morality* (New York: Rowman & Littlefield, 2000).

[205] Part of the life of a squirrel (or any organism) involves reproducing, so some might argue that propagation of the species is the "purpose" of the organism's life. Certainly, the success of past reproduction on the part of earlier generations is required for an organism to exist. But once the organism is alive, its *life* is what it acts to achieve, and reproductive activities may or may not form part of this life. For a more detailed discussion of this issue, see Harry Binswanger,

The Biological Basis of Teleological Concepts (Los Angeles: Ayn Rand Institute Press, 1990), 153–9.

[206] Rand, *Virtue of Selfishness*, 16.

[207] Sam Harris, *The Moral Landscape: How Science Can Determine Human Values* (New York: Free Press, 2011), 12.

[208] Rand, *Virtue of Selfishness*, 25.

[209] Rand, *New Intellectual*, 128.

[210] Ibid., 129.

[211] Rick Warren, *The Purpose Driven Life: What on Earth Am I Here For?* (Grand Rapids, MI: Zondervan, 2002), 30.

[212] For a powerful argument against altruism, see Peter Schwartz, *In Defense of Selfishness: Why the Code of Self-Sacrifice is Unjust and Destructive* (New York: Palgrave MacMillan, 2015)

[213] Several of the points that follow are from Alan Germani, "The Mystical Ethics of the New Atheists," *The Objective Standard*, Fall 2008.

[214] Christopher Hitchens, *God is not Great: How Religion Poisons Everything* (New York: Hachette, 2007), 256

[215] Richard Dawkins, *The God Delusion* (New York: Houghton Mifflin, 2006), 262–72

[216] Sam Harris, *The End of Faith: Religion, Terror, and the Future of Reason* (New York: W. W. Norton, 2004), 180–1.

[217] Jacob Bronowski, *The Ascent of Man* (Boston: Little, Brown, and Company, 1973), 19.

[218] Paul S. Martin, "Prehistoric Overkill," in *Pleistocene Extinctions*, ed. P. S. Martin and H. E. Wright, Jr. (New Haven, CT: Yale Univ. Press, 1967).

[219] Andrew Bernstein, *The Capitalist Manifesto: The Historic, Economic and Philosophic Case for Laissez-Faire* (Lanham, MD: University Press of America, 2005), 105–6.

[220] Indur M. Gaklany, "Weather and Safety: The Amazing Decline in Deaths from Extreme Weather in an Era of Global Warming, 1900–2010," Reason Foundation Policy Study 393, Sept. 2011, reason.org/files/deaths_from_extreme_weather_1900_2010.pdf.

[221] The figure of 85% comes from 2018 data: bp.com/content/dam/bp/busi-ness-sites/en/global/corporate/pdfs/energy-economics/statistical-review/bp-stats-review-2019-full-report.pdf

[222] news.gallup.com/poll/1675/most-important-problem.aspx.

[223] Steven Pinker, *Enlightenment Now: The Case for Reason, Science, Humanism, and Progress* (New York: Viking, 2018), 122.

[224] Michael Crichton, "Environmentalism Is A Religion," speech delivered at Commonwealth Club, San Francisco, CA, Sept. 15, 2003.

[225] Freeman Dyson, "The Question of Global Warming," *The New York Review of Books*, June 12, 2008.

[226] Adam Vaughan, "James Lovelock: Environmentalism Has Become a Religion," *The Guardian*, Mar. 30, 2014.

[227] John Muir, *A Thousand-Mile Walk to the Gulf* (New York: Houghton Mifflin, 1916), 122.

[228] D. Petersen, "The Ployboy Interview," *Mother Earth News*, Jan./Feb. 1975, 21.

[229] David M. Graber, "Mother Nature as a Hothouse Flower: 'The End of Nature' by Bill McKibben," *Los Angeles Times*, October 22, 1989.

[230] Todd May, "Would Human Extinction Be a Tragedy?" *New York Times*, December 17, 2018.

[231] At the first Earth Day celebration in 1970, Dr. Kenneth E. F. Watt (professor of evolution and ecology) said: "If present trends continue, the world will be ... eleven degrees colder by the year 2000. This is about twice what it would take to put us in an ice age."

[232] Paul H. Ehrlich, *The Population Bomb* (New York: Ballantine Books, 1975), xi–xii.

[233] Paul H. Ehrlich & Ann Ehrlich, *Population, Resources, Environment: Issues in Human Ecology*, 2nd ed. (San Francisco: W. H. Freeman, 1970/1972).

[234] *Congressional Record*, Jan 19, 1970.

[235] Robert Zubrin, *Merchants of Despair* (New York: Encounter Books, 2012), 117.

[236] Mostafa Tolba, "Ecological Disaster Feared," *The Vancouver Sun*, May 11, 1982.

[237] Bill McKibben, *The End of Nature* (New York: Random House, 1989), 146 & 150.

[238] Paul Ehrlich, *The Machinery of Nature* (New York: Simon & Schuster, 1986), 274.

[239] Alex Epstein, *The Moral Case for Fossil Fuels* (New York: Penguin, 2014), 10.

[240] Ibid., 26–8. One sign of authoritarianism is the phrase "I believe in science," used as a sign of almost religious belief. See Robert Tracinski, "Why I Don't 'Believe' in 'Science'," *The Bulwark*, March 26, 2019, https://thebulwark.com/why-i-dont-believe-in-science/.

[241] Patrick J. Michaels and Paul C. Knappenberger, *Lukewarming: The New Climate Science That Changes Everything* (Washington DC: Cato Institute, 2016), 65–7.

[242] As just one example of a scientist labeled a "denier," consider Patrick J. Michaels and his book *Lukewarming: The New Climate Science That Changes Everything* (Washington DC: Cato Institute, 2016).

[243] Ayn Rand, "The Argument from Intimidation," in *The Virtue of Selfishness* (New York: Signet, 1964).

[244] See Glenn Rickets, "The Roots of Sustainability," *Academic Questions*, 23(1):20–53, March 2010. Even on its face, the term "sustainability" is problematic. According to Alex Epstein, "'Sustainability' is not a good way to think about things. 'Sustainability' implies that our goal should be repetition: that we want to do something that can be repeated over and over and over. But I think of life in terms of evolution or progress. We don't want to do the same thing over and over and over, we want to find better and better ways of doing things."

[245] John Tierney, "The Reign of Recycling," *New York Times Sunday Review*, Oct. 3, 2015.

[246] Michael Munger, "For Most Things, Recycling Harms the Environment," American Institute for Economic Research, Aug. 14, 2019, https://www.aier.org/article/most-things-recycling-harms-environment.

[247] John Tierney, "Recycling is Garbage," *The New York Times Magazine*, Jun. 30, 1996.

[248] Peter Schwartz, "The Philosophy of Privation," in Ayn Rand, *Return of the Primitive: The Anti-Industrial Revolution* (New York: Meridian, 1999); also reprinted in *Why Businessmen Need Philosophy*, ed. D. Ghate & R. R. Ralston (New York: New American Library, 2011).

[249] Marian L. Tupy, "The Totalitarianism of the Environmentalists," Foundation for Economic Education, July 23, 2017, https://fee.org/articles/environmentalism-is-totalitarian/.

[250] Joel Garreau, "Environmentalism as Religion," *The New Atlantis*, Number 28, Summer 2010, 61–74.

[251] See Andrew Dessler, *Introduction to Modern Climate Change*, 2nd ed. (Cambridge: Cambridge University Press, 2016).

[252] https://www.theguardian.com/environment/series/climate-wars-hacked-emails

[253] This has been noted by numerous scientists who are climate skeptics, such as Patrick J. Michaels. See also the recent work of Dr. Mototaka Nakamura: *Confessions of a Climate Scientist: The global warming hypothesis is an unproven hypothesis* (Sept 2019).

[254] As environmentalists such as Michael Shellenberger point out, nuclear energy is actually much safer, scalable, and less land-intensive than wind and

solar energy. Ironically, if environmentalists had not bitterly opposed nuclear energy in the past, the US would not be so reliant on fossil fuels.

[255] Alex Epstein, "Biotech vs. Bioethics: The Technology of Life Meets the Morality of Death," *The Intellectual Activist* 17, no. 7, July 2003.

[256] Douglas Murray, *The Strange Death of Europe: Immigration, Identity, Islam* (London: Bloomsbury, 2017), 222.

[257] Antoine-Nicolas de Condorcet, *Sketch for a Historical Picture of The Progress of the Human Mind*, trans. June Barraclough (Westport, CT: Greenwood Press, 1955), 72.

[258] John W. Draper, *History of the Conflict between Religion and Science*, (1874; repr., New York: Appleton, 1912), vi.

[259] Ibid.

[260] Ibid., 218–9

[261] George L. Burr, "Andrew Dickson White" entry, in *Dictionary of American Biography* vol. X part 2 (New York: Charles Scribner's Sons, 1964), 89.

[262] Andrew D. White, *A History of the Warfare of Science with Theology in Christendom* (New York: Appleton, 1897), Preface.

[263] David C. Lindberg and Ronald L. Numbers, "Beyond War and Peace: A Reappraisal of the Encounter between Christianity and Science," *Perspectives on Science and Christian Faith* 39 (1987), 141.

[264] Jeffrey B. Russell, *Inventing the Flat Earth* (New York: Praeger, 1991).

INDEX

A

Abbāsid, 56, 60, 65
Academie Royale des Sciences, 103
Accademia dei Lincei, 103
Accademia del Cimento, 103
Agatharchus of Samos, 27
Agricola, Georg, 87
Albategni. *See* al-Battāni
Albert the Great, 79
Alberti, 85
Alexandria, 23, 25, 26, 29, 31, 47
algebra, 57
Alhazen. *See* al-Haytham
Almagest, 31–2, 58, 60
altruism, 49, 171, 184
Ambrose, 42–3, 78
American Civil Liberties Union, 150
Anaximander, 27, 128, 138
Anaximenes, 14
Antikythera mechanism, 29–30
Apocalypse, 120, 181
Apollonius of Perga, 26, 28, 31
Aquinas, Thomas, 79–82, 84, 88, 104,
    107, 110
arbitrary, 116–8
Arcetri, 98
Archimedes, 25–6, 29, 175
argument from design, 134, 137–8, 154,
    156, 190

argument from intimidation, 183
Aristarchus of Samos, 28, 89
Aristotle, 18–22, 28, 32, 39, 52, 62, 82–4,
    167–8
    rediscovery of, 76–8
al-Ash'ari, Abu Hasan, 67
Ash'arite, 67–8, 73
atheism, 107, 134, 149, 191
Augustine of Hippo, 43–5, 81, 83, 135
authoritarianism, 43, 72, 182–3
Averroës. *See* Ibn Rushd
Avicenna. *See* Ibn Sīna
axioms, 25, 50–4

B

Babylon. *See* Mesopotamia
Bacon, Francis, 87, 102, 104–5, 126
Barberini, Maffeo, 95
Barbour, Ian, 109
Barnes, Jonathan, 20–1
al-Battāni, 60
Beagle, H.M.S., 136, 142
Beecher, Henry Ward, 148
Behe, Michael, 153–4
Bellarmino, Roberto, 94–6
Bentham, Jeremy, 173
Bible, 6, 57–8, 61
    depiction of God, 110
    Fundamentalism, 148–50
    Galileo's interpretation, 93

ABOUT THE AUTHOR

F REDERICK M. SEILER has an M.A. in the History of Science from Indiana University, and degrees in Electrical Engineering from Carnegie Mellon University and Rensselaer Polytechnic Institute. He has published articles in *The Intellectual Activist* and *The Objective Standard*. He currently works as an engineer in northern Virginia.